T0220869

# NATURAL SECURITY

# NATURAL SECURITY

A Darwinian Approach to a Dangerous World

*Edited by*

RAPHAEL D. SAGARIN
TERENCE TAYLOR

UNIVERSITY OF CALIFORNIA PRESS   Berkeley   Los Angeles   London

University of California Press, one of the most distinguished
university presses in the United States, enriches lives around the
world by advancing scholarship in the humanities, social sciences,
and natural sciences. Its activities are supported by the UC Press
Foundation and by philanthropic contributions from individuals and
institutions. For more information, visit www.ucpress.edu.

University of California Press
Berkeley and Los Angeles, California

University of California Press, Ltd.
London, England

Library of Congress Cataloging-in-Publication Data

Natural security : a Darwinian approach to a dangerous world /
edited by Raphael D. Sagarin, Terence Taylor.
    p.   cm.
  Includes bibliographical references and index.
  ISBN 978-0-520-25347-6 (case : alk. paper)
  1. Adaptation (Biology)   2. Natural selection.   I. Sagarin, Raphael
D.   II. Taylor, Terence.
  QH546.N38 2008
  576.8'2—dc22

                                                    2007032379


10   09   08
10   9   8   7   6   5   4   3   2   1

# CONTENTS

# CONTRIBUTORS

CANDACE S. ALCORTA Department of Anthropology, University of Connecticut, Storrs

SCOTT ATRAN National Center for Scientific Research (CNRS), Paris, France; University of Michigan; and John Jay College of Criminal Justice, New York City

DANIEL T. BLUMSTEIN Department of Ecology and Evolutionary Biology, University of California, Los Angeles

GREGORY P. DIETL Department of Geology and Geophysics, Yale University, New Haven, Connecticut

DOMINIC D. P. JOHNSON Department of Politics, University of Edinburgh, UK

FERENC JORDÁN Collegium Budapest, Institute for Advanced Study, Budapest, Hungary

KEVIN D. LAFFERTY Western Ecological Research Center, United States Geological Survey, Marine Science Institute, University of California, Santa Barbara

ELIZABETH M. P. MADIN Department of Ecology, Evolution, and Marine Biology, University of California, Santa Barbara

JOSHUA S. MADIN National Center for Ecological Analysis and Synthesis, University of California, Santa Barbara, and Department of Biological Sciences, Macquarie University, New South Wales, Australia

ELIZABETH M. PRESCOTT Washington, D.C.

RAPHAEL D. SAGARIN Nicholas Institute for Environmental Policy Solutions, Duke University, Durham, North Carolina

KATHERINE F. SMITH Institute of Ecology, University of Georgia, Athens

RICHARD SOSIS Department of Anthropology, University of Connecticut, Storrs and Department of Sociology and Anthropology, Hebrew University of Jerusalem, Mt. Scopus, Israel

TERENCE TAYLOR Washington, D.C., Bradley A. Thayer, Department of Defense and Strategic Studies, Missouri State University, Fairfax, Virginia

GEERAT J. VERMEIJ Department of Geology, University of California, Davis

LUIS P. VILLARREAL Center for Virus Research, University of California, Irvine

# ACKNOWLEDGMENTS

Some people immediately and without hesitation jumped at the chance to help this project along. For them, I am immensely grateful, for they are the ones who reminded me that I wasn't completely crazy, or that if I was, I was at least in good company. I count the very first person I broached this idea with, Anne Solomon at the Center for Strategic and International Studies, among this lot. It was her enthusiasm and especially her assurances that everyone in the security field was not already mining the biological world for solutions to vexing post-9/11 problems that sent me to pursue the idea further. Donald Kennedy, who I have known since I was a student at Stanford University in the early 1990s, is inevitably who I go to with a new idea of uncertain merit. From delving into the history that allowed E-Z Cheese to slip through early regulations on chlorofluorocarbons, to using a 90-year gambling record to track climate change in Alaska, to the present investigation, "DK" has always been an open ear and often an enthusiastic supporter of my most outlandish projects.

Beyond encouragement in theory, I was fortunate to find supporters who could provide the resources—financial and intellectual—to back up this project. My father-in-law, Chester Crocker, has spent his career thinking about security as U.S. Assistant Secretary for African Affairs in the 1980s, and more recently as Chairman of the Board of the U.S. Institute for Peace. He was incredibly helpful in recommending colleagues, pointing out the latest security literature, and providing background on international relations and global conflict. NCEAS Director Jim Reichman, who was immediately excited about the topic, was good enough to encourage me to reapply even after my first NCEAS Working Group proposal was rejected by a befuddled review committee. NCEAS is fortunate to be able to provide an incredible staff to working groups, and I am grateful to all of them for successfully getting a dozen or so very busy people together in Santa Barbara and supporting their daily needs on three separate occasions.

My working group participants were wonderful colleagues who eagerly engaged in some of the most creative, thought-provoking and eye-opening discussions I have had in my scientific career. Four working group participants, two young and two seasoned veterans, deserve special thanks. Dan Blumstein and Dominic Johnson both jumped immediately and with great determination into the project. They both did so at pivotal points in their developing careers (Dan during his tenure review and Dom while juggling early fellowships at UCLA and Princeton) without so much as a tremor of outward fear at what their tenure committees or potential employers would think. Geerat Vermeij, whose 1987 book *Evolution and Escalation* was one of the few serious treatments of biology and security that I could find while researching my *Foreign Policy* article, was also an immediate collaborator. But beyond his initial enthusiasm, his constant challenges to working group discussions, critical readings of every chapter in this volume, and vast wealth of knowledge of biological organisms both moved our discussions forward and kept them firmly grounded in reality. His awfully clever (and sometimes just awful) puns interjected discussions so frequently that at one point I proposed a "pun jar" that if filled with Geerat's contributions (at a quarter a pun) would fund our research activities indefinitely. Finally, my coeditor Terry Taylor brought a wealth of completely different experience than Geerat's to the table. As Geerat continually interjected biological reality, Terry continually provided political reality, as well as plainly fascinating stories from his many impressive former careers.

I would be remiss to suggest that ours is a group of "yes" men and women who are unequivocal cheerleaders of Darwinian Security. In fact, some of the best supporters of our work have been the people who first expressed strong skepticism at the idea but were gradually won over. Chief among these types is Chuck Crumly, science editor at University of California Press. From a one-line dismissal of my initial email query, to a ten-minute discussion at the Society for Integrative and Comparative Biology conference a few months later, to a dinner in Montreal during the Ecological Society of America conference in summer 2005, Chuck not only kept an open mind to what this book could be about, but provided some critical suggestions derived from his experience in scientific publishing. My best editor, however, is my wife, Rebecca Crocker, and you may wish to thank her personally for sparing you a multitude of unnecessary clauses, convoluted sentences, and bizarre word choices.

I would further like to thank my scientific mentors Chuck Baxter and Steve Gaines who in addition to being great natural historians, have shown me that science is primarily a creative and holistic endeavor. It was they who taught me that by asking an unexpected question to a difficult problem, you may find the answer right in front of you.

*Raphael D. Sagarin*

Part One

# INTRODUCTION

Chapter 1

# THE ORIGINS OF NATURAL SECURITY

RAPHAEL D. SAGARIN

## The History

Disease, resource scarcity, natural disasters, conflicts, and deadly conflict have threatened human societies for thousands of years. But these threats are not unique to humans. In fact, the rest of the biological world has faced them for over 3.5 *billion* years. Biological organisms have developed millions of responses to these threats, as evidenced by the incredible diversity of body forms, behaviors, and other methods of surviving and reproducing. Some of these responses have been wildly successful, others less so. Yet even among the extinct forms that we know of, many enjoyed a tenure on Earth longer than the years that humans have inhabited the planet. There is much that humans can learn from biological organisms about how to maintain security in a hostile environment. Increasingly, biological organisms and their behaviors are being used as guides to understand and improve economics, medicine, computing, robotics, and energy production (Nesse and Williams 1994; Benyus 1997; Vermeij 2004). Strikingly, the very features that allow organisms to survive and reproduce against a wide range of threats have never been fully probed for their ability to improve our own security.

The blueprints for these biological security systems are not classified but are laid out in fossil organisms, in fragments of DNA, and in the observable behaviors of the organisms themselves. The patterns that emerge raise questions that have immediate resonance for security studies. Why did some animals survive the mass extinctions of the past and not others? How does the immune system identify and respond to the multitude of potential pathogens it is faced with? Why did animals run uphill to safety well in advance of the December 2004 tsunami that killed over 230,000 people in the Indian Ocean region?

Ignoring the struggle and survival of biological organisms even as we are asking ourselves what is necessary to survive in a world of 6.5 billion people is an oversight of both security experts and biologists, and a failure of both groups to communicate with one another. This is not surprising, given that career specialization coupled with the requirements of funding agencies and the limited focus of academic departments create institutional barriers to cross-disciplinary work. This remains true despite the increasing use of "cross-disciplinary" and other such catchphrases in the literature of funding agencies and academic departments. From my own perspective, even immediately after 9/11, security concerns were far removed from my professional thinking, which was intensely focused on the lives of invertebrates in Pacific coast tide pools. As an ecologist studying the long-term responses of these organisms to climate warming, I had still not made the connection between how organisms cope with a changing environment and how humans could cope with the changed environment post-9/11.

Breaking down disciplinary barriers often requires a radical departure from one's traditional role. My departure occurred in 2002/3, when after completing my doctorate and postdoctoral work in marine ecology, I took a year off from academia to serve as the Geological Society of America's Congressional Science Fellow in Washington, D.C. My duties there while working on the staff of Congresswoman Hilda L. Solis often strayed far from my scientific background. One day I would be analyzing tax law and on another writing a eulogy for Senator Paul Wellstone. But I kept a naturalist's eye on my surroundings, noting changes in the environment, the movement of resources, and the behaviors of Washington's denizens.

In late 2002, just a year after the 9/11 attacks, it would have been impossible to spend any time in Washington, D.C., without observing the sense of fear and desire for security that pervaded the Capitol Hill ecosystem. Jersey barriers continually emerged overnight like fungus in rings around monuments, museums, and government buildings. Mail arrived, uselessly, months after it was sent, brittle from the radiation treatments it had undergone in some midwestern processing facility. Seasoned Capitol Hill staffers and young interns jumped tensely at any loud noise (which often turned out to be construction from the enormous bunkerlike visitors' center being carved out under the Capitol), and everyone kept portable chemical masks under their desks.

I found one retreat from this tense environment: the recently restored Botanical Gardens at the base of Capitol Hill. For a few glorious months, one could simply stroll into the gardens and relax in the verdant canopy under the glass. But soon enough (I imagine a memo was written, an audit conducted, funds allocated) the expected phalanx of metal detectors and security guards were stationed at the doors to the gardens, and a visit there became one more reminder of our insecurity.

I watched these enhanced security protocols around the Capitol for an entire year with a growing sense of unease. I observed that every day Capitol police checked identification of drivers entering the underground parking lots, and they checked the trunks of the cars, but never under the hood or in the backseat. Likewise, people entering buildings were always screened in the same way. There was clearly *more* security in Washington, but it was never *varied* security. And this is what troubled me. The tidepools I studied on the Pacific coast were different every time I visited them, the result of thousands of interactions between individuals of the same species and different species, multiplied by their responses to myriad changes in the environment itself. Nature is never static, and organisms and lineages of organisms survive in part by maintaining variation in the face of nature's variability. Unvarying routines of the enhanced security apparatus in the United States were clearly not designed to cope with the unpredictable security environment after 9/11.

As I began to think about security from a natural history perspective, answers to emerging security questions appeared everywhere in nature. Why was it that financial markets were crippled after the 9/11 attacks? Because they concentrated all of their infrastructure in one place. Organisms, by contrast, adopt redundant strategies such as having multiple gene copies on DNA or producing multiple offspring. Why were many chemical plants found to be alarmingly vulnerable to attack even years after 9/11? Because they developed in a relatively predator-free environment and thus devoted most of their resources into competition, rather than defense. In the natural world, selection ensures that resources are allocated according to the environment—peacocks don't flourish where predators are abundant. Why was a college student able to easily bring box knives, mock explosives, and other contraband past airport security and onto airplanes months after 9/11? Because our security systems, unlike those in nature, did not incorporate variability and unpredictability. In nature, even at the level of DNA, variability is incorporated (for example, by having multiple templates to code the same amino acids) to thwart attackers and parasites.

These basic, security-minded natural history observations became a short essay on evolutionary approaches to security for the journal *Foreign Policy* in 2003 (Sagarin 2003). In writing the essay, however, I knew that 3.5 billion years of evolution and millions of species held many more secrets than one person could understand. I realized that to move beyond the most basic analogies between natural and societal security systems would require a broad dialogue among the most creative thinkers in both biology and policy. I began to cautiously spread the idea with my biologist colleagues and contacts I had made in the security policy world. Almost universally, the responses I received charted the same trajectory. Initially, there was a dose of skepticism (one close friend didn't respond to the first email I sent, thinking from the tagline "Evolution and Homeland Security" that it must be spam or a virus), but this was inevitably followed by a period of thoughtful contemplation that invariably led to my contact suggesting completely new angles on the topic that I had not originally considered. Finally, my contact would give me two or three names of colleagues who he or she thought needed to be involved in the conversation.

Very soon, this discussion outgrew my email processing ability, and I realized that the conversation needed to become more formalized. Moreover, I now had a loose-knit group of ecologists, psychologists, anthropologists, paleobiologists, and virologists, not to mention security analysts, biowarfare experts, and former spies, all excited to further discuss the nexus between biological and societal survival. The National Center for Ecological Analysis and Synthesis (NCEAS), a National Science Foundation–supported think tank in Santa Barbara, California, that was created to support small, 8- to 15-person working groups aimed at tackling interdisciplinary problems in ecology, proved to be the ideal venue to move the discussion forward.

With NCEAS support, I began the process of bringing together an incredibly diverse group of scholars and practitioners, some with opinions that directly contradicted one another, to constructively engage in a dialogue about a topic on which few of us had any direct professional experience. Serendipitously, just weeks before our first working group meeting, I learned about Open Space Technology (OST), a meeting facilitation system pioneered by Harrison Owen (Owen 1997). OST is remarkably simple. On the morning of the first meeting, I handed out index cards and Sharpies and asked participants to write two or three topics for discussion about which they either had expertise to lead a discussion or were just simply curious to know more about. We then taped these cards to a white board with an empty agenda printed on it and arranged the topics sequentially into time slots in a way that seemed to make sense to the group. From there the discussion started and did not stop for three days. At times we went back to the cards on the impromptu agenda to start another topic, and at other times we abandoned the agenda altogether.

The central characteristic of an OST meeting is that participants, rather than the organizer, are given almost all of the power to design the meeting and its outcome. In this way, OST mimics a basic organizational process in nature, and one that authors of this volume, especially Geerat Vermeij, have promoted as a critical organizing principle for security systems. That is, successful organisms rely on a system of multiple semiautonomous units that sense the environment and devise solutions to environmental problems with only limited central control. Clonal organisms, such as corals, for example, have been wildly successful using this organizational strategy (although climate warming is now taxing the resilience even of the longest-surviving corals).

## The Book

This book further extends this natural model of organization. After three intense working group meetings and numerous discussions in between, there remained many intersections of biological evolution and security that we had barely touched. Thus, rather than assigning topics and regulating them to conform to one central thesis, I asked authors to tackle a question that seemed most interesting to them and to explore how it related to a dozen or so themes that emerged during the course of our working group discussions. The solutions derived by the authors are diverse, both topically and stylistically (Table 1.1). In some cases, an analogy between societal security problems and solutions developed in nature is drawn and used to shed light on strengths or weaknesses of societal security systems. These analogies may come from long-term studies of evolutionary development, or from near-term ecological studies that consider the immediate relationships and behaviors or organisms in nature. Other chapters focus on how actual evolutionary processes, such as the development of the ancestral mind and the emergence of human social structures, are affecting the security environment today. Still other chapters use tools that were developed in the context of evolutionary and ecological studies—such as demographic and epidemic models—to address security problems faced by society.

In some cases, authors provided chapters that reflected a synthesis of their many years of natural history study. Geerat Vermeij, for example, built on insights he developed through years of intense study of changes in the fossil record. Vermeij has bridged his natural history observations with social commentary in his recent book *Nature: An Economic History* (Vermeij 2004), and he is also well versed in studies of conflict as reflected in his 1987 analysis of fossilized arms races, *Evolution and Escalation: an Ecological History of Life* (Vermeij 1987). Daniel Blumstein, who has spent an alarming number of days studying the behaviors of marmots and birds in a list of countries that would make a covert agent blush, has used his insights to create a list of 14 security lessons we can derive from the antipredator behaviors of animals. Ferenc Jordán, who is intensely studying the networked relationships of both social wasps in India and school children in Hungary, has provided a tool box for analyzing the many types of networks that play a key role in security analysis, from subway station maps to terrorist organizations to the internet.

In other cases, authors used this opportunity to completely break out of their usual field to explore a biological-political nexus that had intrigued them. Greg Dietl, a paleobiologist who is interested in viewing

TABLE 1.1. Example Applications of Biological Study to Societal Security Concerns

| Utility | Example Security Questions | Chapters |
|---|---|---|
| Evolutionary patterns as instructive analogy | *Paleobiology* | |
| | What leads to escalation of armaments and defenses? | 6 |
| | What systems of organization function most efficiently for security purposes? | 5 |
| | What types of organisms survive mass extinctions? | 3 |
| Ecological patterns as instructive analogy | *Ecology* | |
| | How are networks of relationships (food webs, social networks) created and how are they vulnerable? | 14 |
| | When are phase shifts likely to occur in natural and human social systems? | 11 |
| | *Epidemiology* | |
| | Can terrorism be fought and prevented in the same way as a disease epidemic? | 12 |
| Evolutionary roots of behavior | *Behavioral Ecology* | |
| | What is the most efficient behavioral response to chronic threats? | 10 |
| | How are false alarm signals separated from true signals? | 7 |
| | What behaviors and beliefs reinforce terrorist ideology? | 9 |
| Evolutionary roots of human behaviors | *Evolutionary Psychology* | |
| | Why did seemingly irrational behaviors (such as suicide bombing) develop? | 8 |
| | Why are religious beliefs held so strongly? | 4 |
| Evolutionarily derived systems as models for security systems | *Immunology, Virology* | |
| | Can human security systems function as adaptive immune systems? | 4 |
| Evolutionary and ecological models adapted for human security challenges | *Mathematical Biology, Fish and Wildlife Management, Epidemiology* | |
| | Under what conditions will terrorist populations grow or be driven to extinction? | 12, 13 |

the evolutionary process from an economic perspective, particularly arms races in nature, took the most ambitious leap, delving full bore into the literature on Darwinian approaches to international relations, which he found lacking in biological detail regarding the conceptualization of selection. His synthesis, presented here, is both an important theoretical advance in the field of international relations and the roots of a new practical way of looking at international conflict.

The role of religion and moral beliefs was compelling to several authors who approach the problem from widely different backgrounds. Bradley Thayer, author of *Darwin and International Relations: On the Evolutionary Origins of War and Ethnic Conflict* (2004), puts an evolutionary lens on the motivating factors and potential solutions to the problem of Islamic fundamentalist terrorism. Scott Atran echoes the importance of religion in motivating fundamentalist terrorism but also provides some surprising findings from interviews with Islamic Jihad leaders and individuals who have committed themselves to suicide operations. Richard Sosis and Candace Alcorta provide another perspective by focusing on four themes of religion that are recurrent across cultures and discussing their evolutionary roots. These themes, which relate to the strength of commitment to a religious group, the role of the supernatural, the separation of religious and secular elements, and the importance of the adolescent phase of life in acceptance of religion, are all windows into how fundamentalist terrorist groups may recruit and function. Luis Villarreal takes a sweeping view of group behaviors, linking the earliest development of replicating organisms and their genetic parasites to the development of sensory organs, altruism, and religious beliefs. Villarreal suggests that strongly held religious and moral beliefs, especially those that are resilient to rational, empirically based counterbeliefs, are a natural result of the long-term evolution of group identity, which in turn has its origins in viral immunity.

Collaborations were important for several of the chapters. Dominic Johnson, a biologist turned evolutionary psychologist and political scientist, teamed up with both ecologist and evolutionary biologist Josh Madin and ecologist Elizabeth Madin to produce two chapters of widely different character. Josh Madin and Johnson attempt to put some numbers into the study of evolutionary security by adapting fisheries population models to the problem of culling insurgencies. This work represents an important doorway into the world of mathematical biology that could be applied to numerous security questions. Cellular autonoma models of the type developed by Simon Levin and colleagues, for example, could be used to analyze the spread of terrorist ideology and highlight resilient antiterrorist strategies. Another ecological question that has been addressed in models relates to paradigm shifts, in which a stable system suddenly collapses. Elizabeth Madin and Johnson turn this line of inquiry on its head by asking why humans, who are well aware of the potential for catastrophic paradigm shifts, still typically fail to change course until after disaster has struck. One area of human affairs where a proactive approach has made some headway (although most public health experts would argue, not enough) is in preventing infectious disease and pandemics. Recently, Paul Stares and Mona Yacoubian (Stares and Yacoubian 2005, 2007) have recommended using the tools and strategies of epidemiology to fight terrorism. In this volume, Kevin Lafferty, an ecologist and expert on diseases in aquatic organisms, teamed with Elizabeth Madin and Katherine Smith, who studies biological invasions and large-scale patterns in ecology (biogeography), to examine in detail how useful this epidemiological analogy to terrorism is and in what cases is can be extended.

The biologists in the group were especially grateful to have security experts to provide a realistic picture of the threats humans in society face. Elizabeth Prescott has provided an overview of the organizational strategies of both private sector and governmental security agencies and where they succeed or fail to follow efficient evolutionary strategies. My coeditor Terry Taylor, who has a distinguished career facing security threats as a soldier, intelligence officer, and weapons inspector, has provided insights from his ongoing work to understand the concept of "living with risk," a phrase that could as easily be applied to biological systems as societies.

## The Goal

This volume is aimed to provide a first sweep of the potential biological inspirations for solving security problems in modern society. Nature has literally millions of security solutions—our work here can only begin to skim the surface. Thus, some readers may be disappointed to find that there is no "silver bullet" here that would instantly solve our security problems if only political leaders would listen. Nor by any means do we intend to provide one biologically based blueprint to overhaul all existing security architectures. Just as there are myriad workable solutions to security problems in nature that developed individually in response to nature's variation and variability, no one natural solution to security problems will work for the diverse and variable society we live in.

It has become clear through developing this project that a number people still view biology and society as somehow separate. An early reviewer from a leading policy journal suggested that the gap between scientists and social commentators, highlighted by C. P. Snow's "two cultures" concept of 1959, was still the dominant paradigm today, and thus there was no hope for this undertaking. Yet it was Snow himself who identified the need to bridge the cultural gap if we are to solve society's problems. Since Snow, scholars especially with a biological background, including E. O. Wilson,

Simon Levin, Geerat Vermeij, Gretchen Daily, and others have made a strong case that barriers between the biological world and human society are more a human construction than a natural reality. In fact, the assumed separation between humans and the rest of nature may reflect the very type of evolved protective "belief system" that Luis Villarreal uncovers in his chapter.

Also, ethical considerations have been raised about our approach to security. Some of this is residual skepticism derived from early and uninformed applications of "social Darwinism," which used the concept of survival of the fittest (itself a misinterpretation of Darwin's ideas) to justify eugenics programs and colonial imperialism. These concerns are largely unrelated to the analyses we conduct here, and the misuse of social Darwinism in general is dealt with well by the scholars mentioned above, especially Vermeij in the opening pages of *Nature: An Economic History* (Vermeij 2004). In other cases, the concerns stem from the fear that biologically inspired security systems will violate ethical norms or values in society. For example, the incredible information-processing ability of the immune system when tracking pathogens might inspire an invasive governmental data-mining program—an anathema to proponents of civil liberties. There is no reason, however, to assume that we must unilaterally apply evolutionary lessons. As with any science-based policy, ethical, political, and economic norms will undoubtedly influence the policy debate.

Of more substantial concern here, then, are the areas where the unique attributes of human society make the application of evolutionary principles either more or less feasible than in nature. Nature has the luxury of being able to test security adaptations over multiple individual organisms and generations. Those organisms that survive and reproduce copies of given traits pass the test, those that die, fail. This creates a conundrum for humans. We can imagine lots of different adaptations to consider, but we do not have the luxury of multiple generations to look retrospectively at how certain security adaptations turned out, nor do we ethically accept the idea of experimentally exposing humans to dangerous situations to test the effectiveness of security measures.

In this regard we must recall that while we are products of natural selection, we are not completely at its mercy. We have luxuries that most organisms do not have. We have the ability to be self-reflective, and we can deliberately adjust our security systems more quickly and more radically than natural selection would allow in most organisms. We can thus utilize lessons from nature as both warnings and as prescriptions.

In this spirit, my colleagues and I have discussed countless species and natural systems from individual molecules to the global ecosystem known as Gaia. Although each of these species and systems are individual solutions to a unique security problem, I find some comfort in the idea that all of these ideas coalesce into four broad and interlocking themes of naturally inspired security:

*Organization.* Highly centralized organizational structures are unable to efficiently sense the environment and respond accordingly. Using multiple independent sensors with limited central control allows a more complete and adaptable assessment and response to security threats.

*Behavior.* Organisms respond to threats to their security in complex, but often in predictable manners. Although human behaviors appear at first blush far more complex than, say, those of marmots, our behaviors are also a product of evolutionary development, the vast bulk of which occurred in a preindustrial, preagricultural, risk-filled environment. Behaviors that defy modern rational explanation often make sense in light of their evolutionary history.

*Environmental awareness.* All organisms operate under the principle that the world is an inherently risky place. Rather than attempting to eliminate the risk, it is more resource efficient to understand the nature of the risk and allocate resources accordingly.

*Timing.* All natural systems change through time and go through processes of origination, growth, and decay; it is critical to understand what stage the various components of a security situation are in so that we can act appropriately.

I will discuss these four themes and how they relate to one another in detail in the final synthetic chapter. This will allow me to draw on specific insights derived from the individual chapters. Until then, I urge the reader to delve into the diverse community of chapters that follow with an open and creative mind. The lessons you take from these sidelong natural histories may be quite different from my own. If you are like most people who have entered this discussion, you will likely draw a new connection between security in nature and security in society. If so, I encourage you to become part of our growing discussion. For too long, our treatment of security has been removed from the environment in which security threats arose. I hope that this volume is the first voyage of an ongoing expedition to discover the natural history of security.

## REFERENCES

Benyus, J. 1997. *Biomimicry: Innovation inspired by nature.* New York: William Morrow and Company.

Nesse, R. M., and G. C. Williams. 1994. *Why we get sick. The new science of Darwinian medicine.* New York, Vintage Books.

Owen, H. 1997. *Open space technology: A user's guide.* San Francisco: Berrett-Koehler Publishers.

Sagarin, R. 2003. Adapt or die: What Charles Darwin can teach Tom Ridge about homeland security. *Foreign Policy* (September/October): 68–69.

Stares, P., and M. Yacoubian. 2005. Terrorism as virus. *Washington Post*, August 23.

Stares, P., and M. Yacoubian. 2007. Rethinking the war on terror: New approaches to conflict prevention and management in the post-9/11 world. In *Leashing the dogs of war: Conflict management in a divided world*, ed. C. A. Crocker, F. O. Hampson, and P. Aall. Washington, DC: United States Institute of Peace: 425–436.

Thayer, B. 2004. *Darwin and international relations.* Lexington: University Press of Kentucky.

Vermeij, G. J. 1987. *Evolution and escalation: An ecological history of life.* Princeton, NJ: Princeton University Press.

Vermeij, G. 2004. *Nature: An economic history.* Princeton, NJ: Princeton University Press.

# Chapter 2

# LIVING WITH RISK

TERENCE TAYLOR

As a general proposition it is fair to argue that through generations organisms survive by adaptation in response to risks to their continued existence rather than by eliminating potentially catastrophic risks entirely. In human societies, where a desire for stability can "freeze" a society in particular political, ideological, or religious structures, patterns of agriculture and states of technical and scientific development—collectively or individually—inhibit necessary responses to climatic and wider environmental risks. This characteristic can also make these societies more vulnerable to the risk of being overwhelmed by an adversarial human society using military or other means. This chapter attempts to compare lessons from evolutionary experience in nature to help understand elements essential to the survival of human societies in relation to risks that are potentially catastrophic. What are the limits to the lessons that can be drawn from the nonhuman world with regard to a response to catastrophic risks, given the measure of freedom of decision that human societies can exercise? On the other hand, in their search for stability along with the constraints of bureaucratic structures, do some human societies contain the seeds of their own destruction through inhibitions on the measures required to adapt? This chapter is descriptive rather than prescriptive and is intended to be considered a backdrop to the chapters that follow.

History provides many examples of societies that have collapsed through their inability to adapt to changed environmental circumstances, or due to vulnerability to an adversarial society. While there are a number of historical examples of societies completely perishing or being eliminated in the era of nation-states, in particular over the past 100 years, the process has been more one of adaptation even though highly violent means were employed on a number of occasions. Cases in the former category have been well described by Jared Diamond, for example, the conquest of Aztec and Inca societies in Central and South America by the Spanish principally in the seventeenth century.[1] In this instance the cause was not simply military-technical superiority but also the vulnerability of the politico-religious structures to the risks they faced, as well as a lack of immunity to hugely debilitating infectious diseases brought in by the invaders. These societies were unable to comprehend the range and nature of the risks they faced and, in a technical sense, were unable to respond even if they did. Even without an external threat, societies have collapsed through a failure to perceive the risk to their survival both from climatic and directly related environmental conditions, as well as risks arising from their exploitation of their natural environment. This phenomenon is also well explained by Jared Diamond in his analysis of the demise of Norse settlements in Greenland, and the survival of the Inuit people in dealing with the climatic risks in the same region.[2] He goes on to describe an example of a society (on Easter Island) that implodes through continuous and, ultimately, destructive exploitation of the agricultural environment. This is perhaps one of the starkest examples where an isolated society, mostly unaffected by others, failed to take action—which was within their knowledge and power—to avert a collapse.

It is tempting to argue that the biological principle of natural selection, at least in the past millennium, does not apply to complex human societies. This view can be reinforced by the analyses of various historians and political scientists. For example, in her work *The March of Folly* historian Barbara Tuchman argues, using a series of historical examples from Graeco-Roman times through to the twentieth century, that top decision makers repeatedly continue courses of action despite the fact that it is obvious that the policies concerned will lead to disastrous outcomes.[3] Is there a factor in the internal dynamics of human societies that inhibits necessary adaptation to survive and be successful in the face of potentially catastrophic risks? Perhaps, as in nature, some societies (as in the case of some organisms) evolve to become so specialized that adaptation in response to risks to survival is not possible, and collapse or destruction becomes inevitable.

The late twentieth century offered a further example with the collapse of the Soviet Union and Warsaw Pact countries, without military conflict, and was (arguably) primarily due to an inability to adapt in order to participate in the global economy. This failure to adapt was most likely caused by the nature of its domestic political structures along with the fundamental advances in the transnational flows of information and ideas. While there was an attempt at adaptation after the ascent to power of Mikhail Gorbachev, more radical change was needed if the core state of the Soviet Union, Russia, was to survive as an intact entity. The speed and pace of the initial transformation surprised both those on the inside and those outside of the former Soviet Union. This perhaps was due to a failure to perceive the fundamental risks to its survival,[4] in particular those risks to a command economy arising from the advances in science and technology and their incorporation into transnational trading (in particular the quantum advance in the availability of person-to-person and group-to-group data and information flows outside the government arena). While in this case there was an adaptation, it was a cathartic process, which continues more than two decades after its beginnings.

The Soviet Union was not alone in grappling with the problem of adapting a command economy in response to the risks posed by the transformation of the global economy and global networking in, inter alia, science, trading, religion, and political ideas. At about the same time, China in the 1980s began to take its own path of adaptation. In 1981 Deng Xiaoping, the then head of the Communist Party, laid out the strategy for China's domestic adaptation by providing the theoretical underpinning for the introduction of a market economy into a socialist state, despite opposition from other members of the senior leadership. For example, the dissolution of the agricultural communes began through a bottom up approach by local leaders, which eventually transformed the rural economy without seriously disputing the leading role of the communist party in national affairs.[5] By the 1990s the market economy was in full swing and had penetrated to most other sectors of the economy. Furthermore the impetus for change was moved from the rural to the urban and industrial economy. The adaptation remains in progress, with a fundamental difference to the Soviet Union in that, up to now, the state has remained intact, although there have been substantial internal disruptions in certain regions. Another major difference was in the use of the bottom-up approach as opposed to the top-down approach (often referred to as *perestroika* [restructuring]) employed by Mikhail Gorbachev and his successors. Interestingly, the most radical reorganizations of biological systems—mass extinction events—have almost exclusively occurred due to bottom-up forces, as noted in our working group and elsewhere by paleontologist Geerat Vermeij.[6]

The cases of Russia and China are both remarkable examples of adaptation through different approaches in dealing with potentially catastrophic risks.[7] The former is more reactive in character, while the latter appears to show a more preemptive approach. It is important to note that the risks impelling adaptation have less to do with the pressures of great power politics and more to do with a rapidly changing scientific, technical, and economic environment. Governments and international governmental organizations are not the leading edge of change and adaptation in this regard. These national and international institutions and structures, for the most part, represent a response to the state of the world and its attendant risks of some 60 years ago, perhaps even longer. They are largely the legacy of "great power" politics. While governments have diminishing control over

forces that are transforming societies in the twenty-first century, their structures and analytical approaches, in general, have not adapted to this reality.

A former Vice-President to the World Bank, Jean-Francois Rischard is
one of few to write authoritatively on this subject. Commenting on the
implications of developments in global trading and advances in biotechnology in particular, he notes that "global issue networks can transcend the
limits imposed by contemporary territorial and hierarchical institutions
that were never constructed to address these inherently global changes."[8]
Rischard's remarks provide particularly stark examples of areas of technical
development where the advances are moving too quickly for even the most
well-intentioned and organized governments to keep up by providing policy guidance and proposing regulation—or even comprehending where
regulation might be needed—in order to mitigate risks. In this regard
Rischard further comments on the necessary bottom-up approach saying,
"[International] networks can define norms for governments, business,
and multilateral institutions."[9] In effect what Rischard is saying is that those
at the leading edge of the scientific and technical advances are best placed
to understand not only the rewards to human society of these advances but
also the risks they pose. If this understanding leads to direct action—for
example, sharing of best practices and building norms for handling and
manipulating dangerous pathogens from nature—it can be effective in mitigating biological risks. More broadly this type of action can be an agent for
the swifter adaptation now needed for societies to deal more effectively with
potentially catastrophic risks, whether they emanate from natural or deliberate and negligent human action.

An important contribution the scientific community can make is by providing, through interaction with the policy-making community, a science-
based risk analysis to help overcome misperceptions and lessen political
bias in understanding, in particular, the risks posed by transnational terrorists and the weapons and techniques they may use. Risk analysis that
does not embed the consideration of the risks arising from this phenomenon in the wider natural and rapidly evolving scientific and technological
environments will be less effective. Nowhere is this more sharply illustrated
than in the policy response to possible biological terrorism.

Given the topic of this volume it is appropriate that much of this chapter should focus primarily on biological risks. However, the considerations
are relevant to other sectors in which there are inherent risks that are
potentially global in their impact. Since the anthrax attack in the United
States in October 2001, there has been a confusing debate about biological safety and biological security. In particular, in the United States a great
deal of resources and money have been put into countering biological terrorism.[10] It has led to the drawing up of a list of pathogenic biological
agents (known as the "select agent" list) on which work and access are

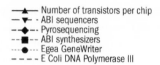

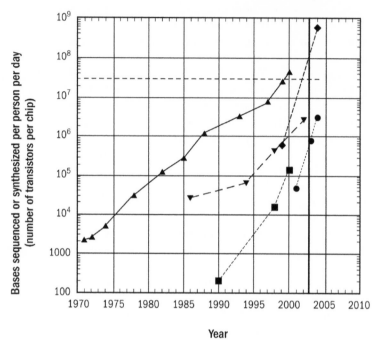

FIGURE 2.1 Advances in DNA synthesis and sequencing: a comparison with Moore's law. From Robert Carlson, "The Pace and Proliferation of Biological Technologies," *Biosecurity and Bioterrorism: Biodefense Strategy, Practice, and Science*, 1: 203–214, 2003.

subject to special legally binding restrictions. Given the advances in the ability to manipulate bacteria and viruses (naturally pathogenic and otherwise), some critics have dubbed this approach as a sort of Maginot Line defense[11] that over time will be ever more readily outflanked as the advances progress, particularly in the area of DNA synthesis and in the emerging field of synthetic genomics.[12] It is estimated that the speed of DNA synthesis has increased more than 500 times between 1990 and 2000 and that "laboratory processes have become more automated and black boxed so that less and less tacit knowledge is required to employ the technologies."[13] Indeed the pace of the advances in biotechnology, impelled by parallel and linked advances in computing power and speed, rival the pace of the advances in computing known as Moore's law,[14] as shown in Figure 2.1, produced by Robert Carlson.

**SPECTRUM OF BIOLOGICAL RISKS**

| Naturally Occurring Diseases | Re-Emerging Infectious Diseases | Unintended Consequences of Research | Laboratory Accidents | Lack of Awareness | Negligence | Deliberate Misuse |
|---|---|---|---|---|---|---|

FIGURE 2.2 From Terence Taylor, "Safeguarding the Advances in the Life Sciences," *EMBO Reports*, 7: S61–S64, 2006, doi.10.1038/sj.embor.7400725.

There is a danger, too, that efforts to defend against biological terrorism—a manifestly infrequent event[15]—could detract from the ever-present risk of infectious disease that kills an estimated 14 million people a year.[16] Even greater concern arises from the possibility of a pandemic of a virulent strain of influenza, such as the 1918 pandemic strain, which is estimated to have killed up to 50 million people.[17] The range of risks illustrated in Figure 2.2 requires an approach that is a "net overall risk assessment" for which new conceptual approaches are needed. Because of the dynamic nature of the biological risk spectrum, a process of regular reevaluation is needed. The U.S. National Academies points to the need to "continually reassess the degree to which scientific advances or current and future 'platforms' hold for being used for weapons purposes."[18] Such an adaptive approach is a challenge for government bureaucracies, particularly where a number of different departments and agencies will be involved, as in the case of biological safety and security. The issue cuts across the boundaries of public health and security in the international field as well, that is to say, the World Health Organization, the UN Security Council, and related instruments such as the Biological and Toxin Weapons Convention and International Health Regulations.

Freeman Dyson (Princeton University) points to the challenge in dealing with the risks arising from the advances in the life sciences in comparing the driving forces behind the advance and dissemination of information technology in the last two decades of the twentieth century. He attributes a key impetus for the advance and spread of the technology to what he describes as its "domestication," in particular, demand from individual consumers. He predicts that in the first decades of the twenty-first century the same will happen with biotechnology. If Dyson is correct, governmental organizations, national and international, will have marginal influence on the direction of the advance and the pace of its dissemination, as was the case with information technology. Dyson continues saying that as we "domesticate(s) the new biotechnology, we are reviving the ancient pre-Darwinian practice of horizontal gene transfer, moving genes easily from microbes to plants and animals, blurring the boundaries between species."[19] While this perspective is one that is looking a number of decades in to the future, it reinforces the importance of drawing on nature in making security assessments in the complex and dynamic risks faced by society, and in particular, the power of bottom-up forces in reorganizing complex systems.

Returning to considerations beyond the field of biology, while there is still a place for the more traditional analyses, it is disappointing to find leading analysts using the perceptions and language associated with great power politics in a way that suggests that their actions are determinant rather than reactive (but as such still important and influential). It is perhaps useful to relate but one example with regard to China. In the closing paragraph of his recent book *The Tragedy of Great Power Politics,* John J. Mearsheimer (University of Chicago) bemoans what he sees as a failure to constrain China's increasing economic power, saying, "So it is not too late for the United States to reverse course and do what it can to slow the rise of China." And further, as if there were really an interstate political process that could hold back the integration of China into the global economy—as well as the dynamic process within China—he predicts, "In fact the structural imperatives of the international system, which are powerful, will probably force the United States to abandon its policy of constructive engagement in the near future."[20] This prediction appears to ignore the complex network of interacting processes, which has more in common with living organisms and their interactions with their environment: continuous, dynamic, and exploiting the feedback from the interactive process.

While it may be tempting to throw in the towel on predictive risk assessment in the face of a very complex and dynamic world and simply revert to a reactive mode—or worse still, stick to anachronistic, but familiar, analytical methods that do not reflect current and future realities—the insights offered by the authors of this volume can enhance and inform risk assessment that can bring results more relevant to current and future challenges. Encouragingly, the true nature of the challenge is becoming recognized even in the traditional international governmental sphere. In a speech in St. Gallen, Switzerland, on 18 November 2006,[21] the UN Secretary General called for leaders in the academic, governmental, and private commercial sectors of the life sciences to build a forum to help address the risks posed by the advances in the life sciences, in order that the enormous benefits these advances bring can be more widely shared. The plan is to operate this forum outside the formal intergovernmental process to produce more responsive results more in tune with the approach to governance suggested by Jean-Francois Rischard, referred to earlier in this chapter and by Sagarin and Jordan in this volume. This approach is more likely to lead to a better understanding of the complex dynamics that underlie the risks faced by human societies. The challenge for the scientific community is to communicate their insights in a way that can be readily understood, explained, and implemented by policy makers. It is unrealistic to expect that the scientific community can deliver, like the human autoimmune system, a methodology to policy and decision makers that could respond to security risks in general. However, by communicating the relevant experience from specific survival strategies developed in the natural world in the face of a complex array of risks, they could make a substantial contribution to national and international security strategies. In this regard, as Geerat Vermeij points out in his chapter, an important lesson from evolutionary biology is that a strategy for countering catastrophic risks does not have to be optimal; adaptation, in time to assure survival and continuing development, is, however, essential.

## NOTES

1. In *Guns, Germs, and Steel: The Fates of Human Societies* (New York: W.W. Norton & Co., 1997, ISBN 0-393-31733-2) Diamond argues that the dominance and technical state of development of European colonial powers was the result, principally, of the resources available to them through geography, climate, and the availability of food and shelter.

2. In *Collapse: How Societies Choose to Fail or Succeed* (New York: Viking, 2005, ISBN 0-670-03337-5) Diamond argues that while societies can witness the catastrophic risk they face, due a variety of reasons in a number of cases that he analyzes, they did not take steps to mitigate the risks that would have enhanced their chances of survival.

3. Barbara W. Tuchman, *The March of Folly: From Troy to Vietnam*, (New York: Random House, 1984 IBSN 0-345-30223-9).

4. Here I am referring to the survival of the Soviet political and economic structures, not the survival of the Russian Federation.

5. See John Gittings, *The Changing Face of China*, (Oxford, UK: Oxford University Press, 2005, IBSN-0-19-280612-2) for a full explanation and analysis of Deng Xiaoping's hugely influential ideas in which Deng is quoted as saying, "Planning and market forces are both ways of controlling economic activity." Deng also promoted export-led growth, which facilitated China's eventual entry into the World Trade Organization on 11 December 2001.

6. Geerat J. Vermeij, *Nature: An Economic History*, (Princeton, NJ: Princeton University Press, 2004, ISBN13: 978-0-691-11527-6).

7. Here I am referring principally, but not exclusively, to risks to the survival of political entities and structures rather than to the lives of citizens.

8. In "Global Issues Networks: Desperate Times Deserve Innovative Measures," *The Washington Quarterly*, (MIT Press for the Center for Strategic and International Studies of Washington DC, Winter 2002/03, 18).

9. Global Issues Networks, 18.

10. The $6.5 billion allocation by Congress in 2005 and 2006 through the Bioshield program is an example.

11. The Maginot Line, named after French minister of defence Andre Maginot, was a line of concrete fortifications tank obstacles, machine gun posts, and other defenses that France constructed along its borders with Germany in the run up to World War II in the light of experience from World War I. The French believed the

fortification would provide time for their army to mobilize in the event of attack and also compensate for numerical weakness. The success of static, defensive combat in World War I was a key influence on French thinking. The fortification system utterly failed to contain the Germans in World War II, and the term is used today to describe any ineffective protection.

12. This is a body of techniques that permits the building of any specified DNA sequence enabling the chemical synthesis of genes or entire genomes.

13. Professor Christopher Chyba (Princeton) in "Biotechnology and the Challenge to Arms Control," *Arms Control Today* 36, no. 8 (2006), 11–17.

14. Named after Gordon Moore, the cofounder of the Intel Corporation.

15. While others have experimented with biological agents, the only recorded biological terrorist event leading to deaths (five) is the anthrax attack in the United States in October 2001. A U.S. National Academies Report, *Biosecurity and the Future of the Life Sciences* (Arms Control Association, Washington DC: National Academies Press, 2006) points to the dangers of the "select agent" approach.

16. The estimate of deaths from infectious disease is drawn from the statistical annex to the WHO's World Health Report 2004. *The World Health Report 2004— Charging History*, (Geneva Switzerland: The World Health Organization, 2004).

17. Estimates of the number of deaths vary, some putting the number a high as 50 million and more. This estimate is drawn from John Barry, *The Great Influenza* (New York: Viking, 2004), 4, in which he cites data by Niall Johnson and Juergen Muelles. Updating the accounts of global mortality of the 1918–1920 Spanish influenza pandemic. *Bulletin of the History of Medicine*, (2002), 76, 105–115.

18. From Recommendation 2b, p. 235, in the National Academy study cited in note 15.

19. In "Make Me a Hipporoo," *New Scientist* 11 (February 2006), 36–39.

20. John J. Mearsheimer, *The Tragedy of Great Power Politics*, (New York: W.W. Norton & Co., 2001, IBSN 0-393-02025-8), 402.

21. The full text of the speech is at www.un.org, Secretary-General's Statement 18 November 2006. For a fuller expression of the Secretary-General's concept of a biological forum see *Uniting against Terrorism: Report of the Secretary General*, A/60/825 (New York: United Nations, 27 April 2006), paragraphs 52 to 57.

Part Two

# LIFE HISTORY AND SECURITY

Chapter 3

# SECURITY, UNPREDICTABILITY, AND EVOLUTION

Policy and the History of Life

GEERAT J. VERMEIJ

Security has been a central concern of human societies and individual people throughout history. The scholars and policy makers who guide strategies against threats from without and within society approach their craft from the perspective of the human-based disciplines of history, political science, and economics. Human nature in this view is seen as both the cause of insecurity and the solution to it. If human nature is unique to our species, as many reflexively assume, looking to the humanities and the human social sciences for understanding and grappling with matters of security makes sense.

Humans are not, however, the only creatures for whom security is essential. In a world sated with competitors, parasites, disease agents, and predators of every description, and in which events beyond anyone's control disrupt resource supplies and threaten death, all organisms and the units in which they are embedded negotiate a never-ending stream of challenges and risks. The persistence of life over 3.5 billion years of Earth history testifies to the abilities of organisms to resist, respond to, or eliminate these uncertainties, and to create insecurity for others. Every creature living today represents an unbroken line of descent from the dawn of life, and each line of descent has persevered against dangers ranging from the quotidian scourges of disease, predation, and weather to events severe enough to precipitate mass extinction and ecosystem collapse.

Persistence, together with the ability to capitalize on and create opportunity for growth and expansion, ultimately arises from adaptations of individuals to commonplace conditions (Vermeij 2004a). These adaptations, representing different approaches to dealing with everyday threats and possibilities, have unintended, long-term consequences for species, groups, and ecosystems. The history of life over the last 3.5 billion years, coupled with the evolutionary principles governing living things and their interactions, provides a rich source of information about how short-term adaptation to garden-variety challenges and risks affect long-term security in the face of unpredicted calamity.

In this essay, I first briefly defend the argument that nonhuman life and its threats can teach us about human travails. I then consider the sources and characteristics of security. Unpredictability emerges as the characteristic for which adequate solutions are most difficult to craft. Building on what we know about the characteristics and history of threats, I next consider classes of solutions to these threats, and explore how each solution affects overall strategies of life. I suggest that the most successful attributes of life's organization—redundancy, flexibility, and diffuse control—are also the characteristics of human social, economic, and political structure that are best suited to cope with unpredictable challenges. The fossil record provides evidence that indirect adaptation to rare, grave, unpredictable crises has become increasingly effective, as indicated by relatively low magnitudes of extinction over the last 200 million years. Finally, I argue that strategies against unpredictable circumstances are best served by diffuse intelligence, which allows alternative emerging hypotheses to be tested, rather than by strong central top-down control by a single authority.

## Human Nature and Lessons from Nonhuman Life

By limiting their scope to human nature and human experience, policy makers unwittingly overlook the insights and evidence uncovered by biologists and paleontologists. Contributing to this parochial perspective is the widely held notion that human nature differs in fundamental ways from the adaptations of other life forms, not least because the problems faced by humans do not apply to other living things. According to this perspective, human uniqueness makes any insights from evolutionary theory and the history of life irrelevant or even misleading.

I contend, however, that humans face the same fundamental problems of competition for locally limiting resources and vulnerability to the vagaries of climate and geological upheaval that other life forms have confronted through the ages. More importantly, the distinctive attributes of human individuals and groups are subject to the same rules of competition and cooperation that have governed the adaptation and evolution of organisms always and everywhere (Vermeij 2004a). Like other species, humans have evolved (often culturally rather than genetically in our species) a multitude of adaptations that enhance our ability to acquire and retain food, territory, energy, mates, money, and a host of other essentials. To be sure, some of these adaptations are uniquely human: complex language, intentional intelligence, altruism toward nonrelatives, global reach, harnessing sources of energy not used by other species, and extraordinary division of labor among genetically distinct individuals and among social groups. Like the unique traits of other species, these and other expressions of human nature permit individuals to make a living and societies to work. Many other human traits are more widely shared with other animals: cooperation, aggression, decision making on the basis of incomplete and inadequate information, status-seeking, and modification of the environment, among others. None of these attributes, whether unique to humans or more broadly distributed in the realm of life, guarantees survival, propagation, access to resources, security, or anything else; but adaptations do enhance individual and often collective performance, and they do imbue their bearers with short-term competitive advantages.

This fundamental identity of human nature with the natures of other life forms opens the door wide to the principles and major findings of disciplines dealing with life's evolution in a challenging world. At the very least, this approach can help us identify solutions that have worked in the past for many forms of life over the long run, as well as point to responses that have proved less effective.

## Characterizing Threats and Security

It is impossible to contemplate ways of enhancing security without carefully defining terms and characterizing the nature of threats. I shall define security as the absence of, or effective adaptation to, agencies that have the capacity to kill individuals, disrupt systems in which individuals live, or impose substantial constraints on activity. Relevant measures of security are the average life expectancy of the affected unit, reflecting persistence; and average performance, measured in units of power, reflecting freedom from constraint. It should be clear that the notions of security and insecurity are relative. Individuals may be secure either because there are no effective threats or because their means of coping with threats are so effective that the threats have been effectively neutralized. By the same token, a living being can be quite insecure if it is poorly adapted, whether the threat is weak or powerful.

Threats are disruptive not only because of the immediate harm they inflict on victims, but because the resulting damage can spread. All biological entities, from the biosphere to the cell, are highly interconnected structures consisting of components that interact with one another. An attack on one component therefore has repercussions for other components and for the larger system as a whole. If the threats are themselves living things, they too form networks, so that communication among them increases the power and reach of danger.

These collateral effects are well known for mutations and extinctions. A mutation, especially the disabling of a regulatory or "switch" gene, affects all proteins and interactions controlled by the gene in question (Carroll 2005). The fewer downstream processes this switch gene affects, the more limited will be the collateral disruption of a mutation in the switch. Similarly, the extinction of a species that carries out a critical function in a community or ecosystem will have major repercussions, because so many other species depend on that function; whereas the extinction of a species that interacts with at most one or two other members of its community will have little effect. Elsewhere I have made the case that when an entire functional group of species, notably food producers dependent on light, is eliminated through a protracted absence of essential nutrients or energy sources, the disruption will propagate throughout the ecosystem and ultimately will bring about the system's collapse (Vermeij 2004b). Invasion of a community by a foreign species, elimination of a top predator, and the addition or subtraction of single species generally can profoundly alter the system but will not normally lead to disintegration, because the productive side of the community is not entirely destroyed (Vermeij 2004b).

In ecosystems, destructive threats propagate by disrupting normal interactions among species; in human societies, threats spread and are magnified by a combination of fear and rapid means of communication. This phenomenon resembles the spread of epidemics; it leads to market manias, the business cycle, and the rapid change in fashions and popular culture. Individuals follow the lead of those members of society who are perceived to be most knowledgeable, powerful, or charismatic; they, in turn, transmit their states of mind to still other members (Ormerod 1998). On the one hand, such rapid spread of fear informs and protects individuals before the actual threat arrives; on the other hand, it magnifies the potential effects of the threat. Our social nature therefore propagates and distorts threats and makes us particularly vulnerable to enemies who initiate threats and who manipulate the means for spreading them.

The greatest challenge to security, and the one most difficult to combat, is uncertainty or unpredictability. Security must ultimately come through adaptation. Phenomena whose timing or location is unpredictable cannot be the direct targets of such adaptation. When circumstances are common enough to occur once or more during the lifetime of an entity, they become predictable, enabling the entity to adapt to them (Hutchinson 1953). An adaptation is nothing more than a hypothesis about predictable aspects of an entity's environment, including dangers and risks (Sterrer 1992). In traditional evolutionary biology, adaptation—whether by gene-based natural selection, physiological or behavioral response, culture, or even manipulation—is viewed as a short-term fit between living things and their current circumstances. This perspective led observers such as Gould (1985, 2002) to argue that persistence through rare, unpredictable events depends more on luck than on effective adaptation. According to this view, adaptation is unimportant in the long-term evolutionary patterns of history, because the rare, unpredictable events to which organisms cannot adapt erase and overwhelm adaptations to such everyday threats as enemies. Life lurches through history from one devastating crisis to another without accumulating adaptive know-how and without apparent improvements in constructing hypotheses of its environments.

But this view ignores unintended consequences—adaptive pathways to solve one problem influence the options for dealing with others—and it fails to take into account the possibility that short-term solutions to commonplace threats predispose their creators to construct means of coping with wholly novel challenges. The history of life—with its extraordinary diversity of adaptive solutions, independent evolutionary lineages, varied and changing circumstances, and uncertainties—offers perhaps the most prolific source of data from which we can draw lessons about how to deal with threats to our own security. Most solutions have been tested many times by many lineages over very long intervals of time, including rare events. It is therefore possible to gain a detailed, quantitative understanding of the efficacy of almost all conceivable solutions over a very wide range of time scales and spatial scales, from the ephemeral and local to the nearly timeless and global.

In order to evaluate strategies for coping with risks in general and with unpredictability in particular, it is important to identify what these risks are and how they affect entities and the systems in which those entities reside, and to understand how adaptation to one type of challenge affects vulnerability and response to other less common or more severe risks. An essential element of this endeavor is the reality that risks and the entities they affect are contingent on economic (or social) interactions; that is, neither risks nor entities can be considered in isolation. The full internal and external context of interaction among components, interactions at different levels of an economic hierarchy, and interactions among economic units above that of the affected entity must be considered in order to craft effective ways of managing unpredictable risks.

## What Makes Phenomena Unpredictable?

Because all living entities, including their parts and the larger ecosystems in which these entities are embedded, are adapted, any change in their external environment or internal conditions will be disruptive. If the disruptions are mild enough or common enough, entities can adapt to them; but if the disruptions are severe (causing widespread death) or rare or both, they will be perceived as novel. Examples of such disruptions include (1) the introduction of a predator, competitor, or disease agent whose power, mode of action, or breadth of effect exceeds any that were previously experienced by victims and that were incorporated in the victims' adaptive hypotheses; (2) a sudden physical catastrophe, such as a collision of Earth with an extra-terrestrial body, unprecedented and large-scale flooding, volcanism, gas releases, radiation, novel poisons, and habitat alteration of life; (3) a sudden catastrophic collapse of the resource base, brought on by disruptions of either or both of the first two categories. In the history of life, these unpredictable disruptions often result in extinction. At the largest scale of disruption, typically involving a physical catastrophe and ecological collapse, extinctions occur worldwide and are so numerous that geologists use these times of mass extinction as time boundaries between ages.

Even the most devastating of these crises, however, did not come close to wiping out life on Earth. Individuals and species—even ecosystems—survived the mass extinctions; they regrouped, proliferated, adapted, and created whole new ecosystems. New adaptations were built on historically preserved traits, including adaptations to bygone circumstances as well as the attributes that enabled the ancestors of the new players to survive the great calamities. New communities and new ecosystems formed from old remnants, either in situ survivors or invaders from refuges where the agencies of death were muted in their effects.

## Characterizing Solutions

There is a staggering diversity of adaptations to an equally dizzying array of threats and challenges. Overwhelming and fascinating as this diversity is, the solutions organisms have evolutionarily devised are variations on just a few broadly applicable themes. We can learn lessons from the adaptive history of life if we can extract these broad themes from the more particular variations.

I identify seven overlapping ways for averting, neutralizing, blunting, or eliminating unpredictable threats:

1. Tolerance through passive resistance (or isolation from the threatening agency through suspended animation in a protective device);
2. Active engagement to disable or eliminate the threatening agency through the use of force;
3. Collective observation or intelligence and other means that increase the affected system's power and longevity, so that the threats become effectively predictable;
4. Unpredictable behavior, preventing threatening agents from constructing a hypothesis about the potential victim;
5. Quarantine and starvation, isolating the threat by starving it of resources or reducing the means by which the threat can be spread to other parts of a system;

6. Redundancy, ensuring that function is maintained even as parts of the system are damaged; and
7. Adaptability, a flexible organization of semiautonomous components under weak central control, enabling rapid and effective responses to unforeseen circumstances

Tolerance is usually a passive response to threats. In the presence of danger, individuals or groups can insulate themselves from the outside with a more or less impervious barrier. By curtailing activity and metabolic expenditures, individuals may enter a state of suspended animation, remain undetected by an enemy, or wait until the crisis passes. Variations on the passive theme include protection by or existence as shells, tubes, lodges, houses, burrows, hives, cocoons, seed coats, eggshells, cysts, tubers, bulbs, spores, carapaces, clothing, toxic exteriors, and nests, among others. If food becomes temporarily scarce, or conditions become inclement, the passively defended organism can hold out for periods ranging from hours to months or even years, depending on the permeability of the cover and on the extent to which the machinery of life can be shut down.

Insofar as tolerance implies inactivity, passive adaptation is inconsistent with a high-powered, competitively superior ecological role. This is true likewise of such antipredatory measures as small body size, cryptic coloration, and silence, all of which are adaptations making prey difficult for enemies to detect. Such adaptations are highly effective, but in their pure form they relegate their bearers to an ecologically marginal, if safe, existence. However, if the passive state is temporary—that is, if it can be induced or invoked only when individuals are threatened, and abandoned when conditions are safe—then reliance on impregnable passive defenses need not interfere unduly with the pursuit of power.

Other variations on the theme of tolerance are much less ecologically constraining. These include large body size, maintenance of a homogeneous internal environment in the face of fluctuations outside the body, and the ability to sustain and repair substantial damage. These capacities require large immediate outlays of energy resources and make possible freedom of movement and other activities even though enemies are present. Again, this form of tolerance can be highly effective against enemies and against brief interruptions in energy sources, but in the long run it is incompatible with resistance to protracted, rare calamities, for which only a major reduction in metabolism coupled with protection behind impermeable barriers will suffice.

The second theme includes active defenses, which eliminate, disable, or blunt sources of harm and uncertainty through immediate expenditures of energy. Active responses against enemies include escape, pursuit, aggression, and the production of induced responses or other means to limit and repair damage. All are contingent on the ability to sense and to respond to danger quickly, meaning an effective integration of sensory and motor systems and effective internal communication among parts. The high metabolic rates associated with active defense can be sustained only in a regime of high availability, predictability, and accessibility of energy resources. Active defense is part of what makes individuals competitive dominants and major power brokers in ecosystems.

Predictability of a phenomenon permits an observer to propose and test hypotheses about that phenomenon. In other words, it enables an entity exposed to the phenomenon to become adapted to the challenges the phenomenon brings with it. It is crucial to realize, however, that predictability is determined both by the occurrence of the phenomenon itself and by characteristics of the observer or the affected party. With a longer period of observation and a greater spatial coverage of observations, even quite rare and infrequent phenomena can be made predictable. For observers this means a longer life span and a greater individual or collective intelligence, as well as the ability to retain, filter, and integrate data. Like active defense, these capabilities require high metabolic outlays and therefore a large, predictable supply of accessible resources.

If predictability makes adaptation possible, unpredictability prevents it. A highly effective means of coping with a living opponent is to challenge it in unpredictable ways. This requires sophisticated abilities to randomize those aspects of the enemy's environment that harm the opponent. Such abilities entail processing and modifying large amounts of information and communicating that information (or misinformation) to the enemy. Examples of unpredictable defenses in the animal kingdom take many forms. Many butterflies escape from birds by executing highly erratic movements involving rapid changes in the direction of flight and sending confusing signals. Many predators form search images by which they can learn to recognize profitable prey. The prey may thwart the formation of such visual hypotheses by displaying great variation in color, behavior, or other attributes that the predator can observe. Variations of this kind can be displayed by individuals at different times, or by many individuals in a population at the same time. Many plants and animals have evolved extremely long intervals between the most vulnerable phases of the life cycle. Periodical cicadas in the eastern and central United States, for example, have an essentially unprotected brief flying adult phase that appears every 13 or 17 years, an interval longer than the life span of most of the likely predators of these insects. Many bamboos and some tropical forest trees are characterized by times of mast flowering. So many flowers are produced for such a brief interval after such a long period without flowers that enemies attacking flowers, fruits, or seeds might be unable to predict when resources become available.

An interesting form of unpredictability that may offer some species protection from enemies is low population density. If a species is rare, individuals often occur in mutual isolation, so that specialized enemies may be unable to find them or to recognize them as desirable when found. In tropical rain forests, for example, damage and mortality of trees due to herbivorous insects and fungal pests is reduced if plants live far from neighbors (Janzen 1970). For these plants, it therefore pays to be rare. This option is unavailable to species that depend on wind pollination, which is effective only when plants of the same species grow close to each other; but it is open to species that are pollinated by specialized, wide-ranging bees, birds, or bats. In an environment such as the tropical rain forest, where host-specific pests are overwhelmingly important, this kind of unpredictability—made possible by an evolved reliance on mobile animal pollinators and seed dispersers—may represent an essential component of defense and survival for many species (Raven 1977; Regal 1977; Leigh et al. 2004). On the negative side, such rarity would become a liability if the animal helpers disappeared; and it is inconsistent with an ecologically dominant role. Rarity also exposes the population as a whole to the threat of extinction because, as noted below, there is insufficient redundancy in numbers to cushion the population against lethal disruptions.

Threats are particularly destabilizing in a system when many components of that system are directly or indirectly affected by the threat. Isolation—reducing the links between threat and members of the system, or limiting connections among members—is key to limiting the damage wrought by the disturbing agency. This is the principle behind quarantines, which effectively limit the spread of infectious disease. It is also one of the rationales for restricting economic monopolies and totalitarian governments. Although such organizations are highly effective in spreading good through a tightly integrated system, they can also inflict grave damage, for no compensatory mechanisms are in place to limit the spread of the harm. As discussed above under tolerance, isolation of an individual from harmful agencies is often passive; but isolation of a threat from the rest of a system is a collective effort requiring the same capabilities that render an unpredictable phenomenon predictable.

A related organizational trait that limits damage is redundancy. Even if part of the system fails, functionally identical parts that remain unaffected continue to maintain the system and to prevent its collapse. As noted by Carroll (2005), almost all living systems are modular in construction, being composed of multiple copies of parts that fulfill complementary functions. Redundancy provides stability and robustness in the face of change; it creates a kind of collective tolerance. Redundancy at the level of populations means that populations consisting of many individuals are less susceptible to extinction than are small populations. Multiple copies of genes provide some insurance against lethal mutations. Economic stability against large-scale regional crop failure is provided by growing the same crops in many parts of the world. The same principle applies to ecosystem services, manufacture of critical machinery, construction of roads and other infrastructure related to transportation and communication, military bases, fuel production, and so on. Having a particular function performed by only one component or a small number exposes a system to catastrophic collapse when that component is disabled.

If large numbers of individuals provide a population with a certain collective resistance to large-scale disturbances, small numbers should have the opposite effect. It has long been known that some adaptations with substantial short-term benefits to individuals are harmful in the long run because they are linked with small population size and therefore with a greater vulnerability to environmental change. Large body size, for example, is usually associated with competitive superiority, antipredatory defense, and high reproductive output; but at the population level it is usually accompanied by small numbers of widely scattered individuals. For mammals and probably most other vertebrates, large body size is statistically linked to a high probability of extinction during crises (Van Valen 1975). In particular, large-bodied mammals that are specialized as predators of large prey belong to species with shorter geological life spans than contemporaneous species with broader diets (Van Valkenburgh et al. 2004). Not only do the large top predators have smaller populations, but the resources on which they depend also come in large, scattered packages. The protections offered by redundancy are therefore steadily reduced as animals grow larger, as population numbers and densities decline, and as critical resources undergo similar trends. To compensate for these effects, large-bodied top predators could (1) curtail metabolic activity during times of scarcity, (2) store enough food to survive lean times, (3) broaden the diet to include smaller or more common items, and (4) establish very wide ranges that include some environments or regions not subject to disturbance (Vermeij 2004a).

In the name of efficiency, many economists seek to reduce redundancy. This view rests on Ricardo's principle of comparative advantage, according to which production of a given commodity should occur only in locations where the cost of production is lower than anywhere else. This view is more widespread than ever as trade barriers between nations are coming down. Many regions suitable for crops such as apples, rice, wheat, potatoes, and sugar are being taken out of production because the subsidies necessary to keep production financially viable are being eliminated or reduced. This may make short-term economic sense, but it is a bad long-term policy in that it reduces redundancy and makes the whole world (and many regions in it) more susceptible to regional collapses.

Finally, and most generally, living entities must be adaptable. If threats cannot be predicted or isolated, the affected entity must have the capacity to respond quickly and appropriately to potentially harmful change. This requires an organization of semiautonomous parts under weak central authority. Biological entities ranging from cells to individuals and ecosystems are organized in this way; so are most human social structures. In a cell, the nucleus acts as the weak central authority. For an individual animal, this function is performed by a brain or the central nervous system, which receives inputs from other parts of the body and then coordinates suitable responses. In ecosystems, authority in both an ecological and a longer-term evolutionary sense is vested in foundation species—trees in forests, corals on reefs, barnacles and mussels on wave-exposed rocky seashores, and other species that provide three-dimensional structure for other members of the system—and keystone species, high-powered top predators or herbivores that exercise ecological control over energy flow in the system and evolutionary control by imposing intense selection on the characteristics and distribution of other species (Vermeij 2004a). Flexibility, rapid communication among parts, substantial local control, and limitations on the power of dominant parts that exercise central authority promote the formation and testing of multiple hypotheses and the blunting of harmful agencies.

The ability to generate many potential solutions through combinatorial processes is a key element of adaptability. The vertebrate immune system must contend with a huge range of pathogens. The limited number of genes is unable to fashion heritable defenses against all of these pathogens and cannot cope at all with novel agents; but they can code for subsystems that then create vast numbers of combinations of simple molecular components. With its billions of neural connections, the brain similarly depends on combinatorial principles. Carroll (2005) has similarly emphasized the role of combinatorics in the developing embryo and notes that the general translation of genes into proteins and larger structures involves value-added information provided by combinatorial power.

## Conflicts and Synergies

This survey of strategies for surviving rare, severe challenges reveals three important principles that relate short-term benefits to long-term effects. First, long-term strategies emerge as incidental consequences of adaptation to commonplace phenomena; they are therefore indirect adaptations to infrequent or novel circumstances, which therefore become effectively predictable. Second, all species on Earth possess, or are derived from ancestors that possessed, traits that enable individuals to survive unforeseen conditions. Third, some traits that confer short-term gains in power and influence are inconsistent with long-term strategies of survival.

Studies of mass extinction indicate that all species surviving the great crises that have visited the biosphere from time to time have at least one stage of the life cycle in which the passive-tolerance syndrome is expressed (Vermeij 2004b). Even the metabolically active, warm-blooded birds and mammals that survived the end-Cretaceous mass extinction 65 million years ago, and the rapidly metabolizing land vertebrates that successfully negotiated the end-Permian crisis 250 million years ago, likely had the capacity to enter a state of torpor, aestivation, or hibernation, during which resource requirements are greatly reduced as the metabolic machinery is partially shut down. Such states are commonly observed among rodents, hummingbirds, and lemurs, among many other mammals and birds. Recent physiological work (Blackstone et al. 2005) shows that the presence of hydrogen sulfide, a compound abundant during the oxygen-poor end-Permian crisis, elicits metabolic torpor in mice.

The state of suspended animation that enabled so many individuals to endure the extraordinary conditions prevalent during times of mass extinction cannot be considered an adaptation to those circumstances but is, instead, an accidental if welcome consequence of adaptation to more commonplace phenomena, notably attacks by predators and other enemies. Because the traits associated with suspended animation are part of the legacy of all species on Earth, it is fair to say that rare events have become predictable in an indirect way. This is due entirely to the general applicability of passive resistance: the adaptations of everyday phenomena work even under conditions very different from those in which the traits conferring resistance were first favored.

The organizational properties that enable biological entities to cope with unpredictable circumstances may likewise have originated as adaptations to everyday problems, but they more directly transform unpredictable phenomena to predictable ones. They do so by cooperation, creating multiple novel combinations of preexisting components, preventing threats from spreading, or creating larger biological units that have a longer life span and therefore the means to retain and accumulate information about rare events. Redundancy and adaptability emerge as modules multiply, cooperate, and forge larger stable evolutionary units.

## The Historical Record

Having identified and examined the long-term consequences of the ways in which living things deal with everyday challenges and with unforeseen circumstances, what can we say about the historical record of these different solutions? I propose that life in the aggregate has become more adept at resisting and responding to unpredictable challenges.

When I began to travel widely and to observe how shell-bearing snails and clams on the world's shores make a living, I observed that these molluscs must cope with a large cast of predators that swallow, forcibly enter, drill, pound, cut, or crush the shells of their prey in order to gain access to the edible soft parts within. Not surprisingly, molluscs evolved a host of defenses against their enemies. Many of these involved armor—a small shell opening, a door or other device to restrict the opening, strengthening ribs and knobs, spines and frills to make the shell larger and therefore harder to manipulate, a thicker shell wall, and a shape enabling the soft parts to be withdrawn deeply into the protective shell—but other defenses emphasized toxicity, speed of escape, and even aggression against enemies. Still other lineages, in which these defenses failed to evolve, became specialized to live in places where enemies hunt less effectively or are absent altogether; they came to live under large boulders, burrow far below the surface of sand or mud bottoms, parasitize the bodies of well-defended hosts, or occupy dark caves. All of these specializations—from armor to escape to aggression to life in enemy-free refuges—are most spectacularly expressed in warm, productive regions, where the power of enemies is least constrained by resource supply and accessibility.

Interest in fossils led me to examine enemy-related adaptive syndromes in ancient molluscs. Few of the specializations in shell architecture related to armor and escape show up in molluscs living before 230 million years ago, and all the specializations evolved in multiple lineages since that time. Moreover, the incidence and degree of expression of nearly every type of defense have increased among molluscs over the course of the 540-million-year history of the group. They did so, I suggest, because the enemies of molluscs became increasingly powerful and because they evolutionarily "explored" an increasingly wide variety of tactics to detect, pursue, subdue, and consume prey. The predators, in turn, were competing with each other and were adapting to their own enemies. The entire process—an episodic series of events involving within-lineage evolution as well as the replacement through time of unrelated lineages—is a kind of enemy-directed arms race, or escalation (Vermeij 1987, 2004a).

This tug-of-war between molluscs and their enemies is only a small part of a much more general process of escalation between enemies and their victims. It is a competition-driven phenomenon of change that affects all economic systems. Its rate and extent are determined by resources, which themselves are under the control of organisms. Over time, competitive dominants create strong selection among their prey and determine where and when victims can make a living. Over time, both offensive and defensive weaponry have increased in power. Not every lineage can always adapt to the latest weapons of predators, and those lineages that cannot keep up are ecologically marginalized (though rarely extinguished outright). Because no offense and no defense can ever be made invulnerable, there can never be a perfect defensive strategy or a perfect offensive tactic. Opponents can always, in principle, find ways to overcome part of the other side's weaponry.

Although enemy-related evolution is almost always by adaptation to predictable challenges, the patterns of escalation between enemies and victims are relevant to the history of strategies for coping with unpredictable circumstances. Both passive resistance and active engagement with enemies have increased in frequency and expression over time within any given environment. The ability of many entities to resist or to engage with novel circumstances may likewise have increased if enemy-related adaptations are applicable to an array of challenges much greater than the specific situations under which the traits arose.

The history of extinction offers some support for this claim. Although extinction of single-celled taxa such as bacteria and amoebae may always have been rare (Knoll and Bambach 2000), perhaps because these organisms tolerate extremes well and because they are exceedingly widely distributed, episodes of extinction have been a prominent feature in the history of life ever since complex animal life emerged some 600 million years ago or a little earlier. Extinction provides evidence of lack of adaptation and therefore reflects phenomena that for many species during times of crisis are effectively unpredictable. Although Bambach and colleagues (2004) were unable to detect a consistent decrease in the magnitude of genus-level extinction of marine animals over the course of the last 550 million years, their data indicate very high magnitudes for the first 80 million years of this interval and generally low magnitudes over the last 200 million years (barring the high-magnitude end-Cretaceous extinction at 65 million years ago). This pattern means that organisms have collectively experienced the last 200 million years of Earth history as being more predictable, either because life forms have gotten better at resisting or responding to unpredictable challenges or because even rare phenomena have become predictable. The two interpretations are not, of course, alternatives. The fact that many lineages incorporate elements of both passive and active enemy-related adaptation may mean that the more recently evolved complex entities have cumulatively "learned" about, and therefore "adapted to," even the rarest and most severe circumstances that in early times would have wiped out a larger proportion of the biota.

## Implications for Policy

Before considering policy, it is useful to lay bare the major facts revealed by the fossil record of life and by the principles governing the ways in which living entities interact and evolve. I distill the findings into six points:

1. Threats from lethal, unpredictable challenges will never go away, and no adaptations to them can be perfect or completely effective. It is

incorrect, misleading, and dangerous to claim that a "war on terror" can be definitively won. There will always be diseases, competitors, terrorists, and other exigencies of every kind. Even as old foes are conquered, new ones will arise and adapt. The history of life is a never-ending struggle of life against enemies and other threats, a struggle that has gone on for some 3.5 billion years.

2. The struggle against threats, including wholly novel ones, has been successful for many lineages of life, but not without cost. Adaptations to challenges require substantial allocations of material resources and time. These allocations constrain other competing functions unless the total energy budget can increase, an option available only when resources are sufficiently plentiful, accessible, and predictable.

3. Passive resistance or tolerance, especially if it involves metabolic shutdown, is highly effective in dealing with a great variety of unpredictable threats, but it is often inconsistent with activity and with the exercise of power and influence. Entities employing such strategies enjoy long life spans but tend to be restricted to an ecologically marginal existence.

4. Exclusively active means of meeting unpredictable challenges expose users to the risk of ecological (or economic) collapse because they are dependent on a continuously prolific and predictable supply of resources. Entities employing such strategies exercise great economic power and influence but typically have short life spans. Active strategies must therefore always be complemented by passive resistance and tolerance if power is to be maintained for long periods.

5. Organizational traits, especially redundancy and modular structures in which semiautonomous components exercising diffuse control are integrated into a larger whole and are themselves coordinated by a relatively weak central authority, provide the most flexible, adaptable, and reliable means of making unpredictable challenges predictable and of responding effectively to those threats. These properties are emergent features of interacting systems composed of repeated as well as functionally differentiated parts. They are therefore collective properties that make possible the transformation of unpredictable, unknowable events into phenomena for which testable hypotheses (that is, adaptations) can be constructed.

6. The history of life indicates that adaptation to both everyday and unpredictable circumstances has improved over the course of time. Strategies initiated as adaptations to predictable events have over time enabled life in the aggregate to construct hypotheses about unpredictable phenomena.

The principal implication I derive from these truths is that no capability should be invested in single units, or depend on single components of infrastructure in a system. This principle applies to gathering information about potential threats, interpreting and evaluating that information, formulating policy, growing food crops, extracting natural resources including fuels, conducting research on emerging diseases, training and other military activities, manufacture of essential machinery and other industrial products, coordinating communication and transport, and economic and political control. These functions should be distributed among members and among sites, not unduly concentrated so that all power rests in the hands of a tiny oligarchy. Diffuse responsibility prevents errors made by one party from crippling the entire system and allows for the testing of competing adaptive hypotheses.

In human affairs, as in the affairs of nonhuman life, there has always been a tension between top-down control exercised by a powerful executive and more distributed controls invested in several bodies. As an observer rather than a maker of policy, I have the impression that governments tend toward a more centralized, more totalitarian, and therefore less adaptable structure when they perceive threats, real or imagined. The seductive rationale for such concentration of power is surely that "good" policies—those that preserve sovereignty and protect citizens (or at least those in power)—can be crafted and implemented quickly in a streamlined, strongly top-down power structure, in which opposition is kept down, checks and balances are eroded, and data conflicting with established policy are ignored, denied, or misrepresented. More distributed power is viewed as inefficient, leading to slow and perhaps inconsistent response. Critically, the top-down power structure works only when its policies are indeed the "right" ones; it can fail catastrophically and irreversibly if errors are made, as will almost always be the case. Error, whether it be a mutation or a faulty decision, characterizes all economic systems. It can be reduced in its frequency and in its effects, particularly in a system characterized by modularity and redundancy, but it cannot be eliminated.

The systems I advocate, and the ones that living things appear to have found most adaptable in the face of common as well as novel threats over the long sweep of Earth history, emphasize a kind of diffuse or collective intelligence distributed among many players in many places, rather than an omniscient, omnipotent, totalitarian structure with one authority at the helm. The hypotheses that these systems construct and evaluate are able to collect evidence from many sources. Continual incorporation of new data and testing of strategies in the face of changing circumstances will not guarantee security, but systems built on the economic principles of evolution and on a model of the scientific method have had an enviable track record through turbulent times. We can only hope that those who wield power and carry out policies to enhance security will heed the evolutionary lessons of history.

## REFERENCES

Bambach, R. K., A. H. Knoll, and S. C. Wang. 2004. Origination, extinction, and mass depletions of marine diversity. *Paleobiology* 30: 522–542.

Blackstone, E., M. Morrison, and M. B. Roth. 2005. $H_2S$ induces a suspended animation–like state in mice. *Science* 308: 518.

Carroll, S. B. 2005. *Endless forms most beautiful: The new science of evo devo and the making of the animal kingdom.* New York: Norton.

Gould, S. J. 1985. The paradox of the fist tier: An agenda for paleobiology. *Paleobiology* 11: 1–12.

Gould, S. J. 2002. *The structure of evolutionary theory.* Cambridge: Belknap Press of Harvard University.

Hutchinson, G. E. 1953. The concept of pattern in ecology. *Proceedings of the Academy of Natural Sciences of Philadelphia* 105: 1–12.

Janzen, D. H. 1970. Herbivores and the number of tree species in tropical forests. *American Naturalist* 104: 50 1–528.

Knoll, A. H., and R. K. Bambach. 2000. Directionality in the history of life: Diffusion from the left wall or repeated scaling of the right? *Paleobiology* 26 (Supplement to no.4): 1–14.

Leigh, E. G. Jr., P. Davidar, C. W. Dick, J.-P. Puyravaud, J. Terborgh, H. ter Steege, and S. J. Wright. 2004. Why do some tropical forests have so many species of trees? *Biotropica* 36: 447–473.

Ormerod, P. 1998. *Butterfly economics: A new general theory of social and economic behavior.* New York: Pantheon.

Raven, P. H. 1977. A suggestion concerning the Cretaceous rise to dominance of the angiosperms. *Evolution* 31: 451–452.

Regal, P. J. 1977. Ecology and evolution of flowering plant dominance. *Science* 196: 622–629.

Sterrer, W. 1992. Prometheus and Proteus: The creative, unpredicatle individual in evolution. *Evolution and Cognition* 1: 101–129.

Van Valen, L. 1975. Group selection, sex, and fossils. *Evolution* 29: 87–93.

Van Valkenburgh, B., X. Wang, and J. Damuth. 2004. Cope's Rule, hypercarnivory, and extinction in North American canids. *Science* 306: 101–104.

Vermeij, G. J. 1987. *Evolution and escalation: An ecological history of life.* Princeton, NJ: Princeton University Press.

Vermeij, G. J. 2004a. *Nature: An economic history.* Princeton, NJ: Princeton University Press.

Vermeij, G. J. 2004b. Ecological avalanches and the two kinds of extinction. *Evolutionary Ecology Research* 6: 315–337.

Chapter 4

# FROM BACTERIA TO BELIEF

Immunity and Security

LUIS P. VILLARREAL

The security and stability of a nation, group, or people can be considered as closely analogous to the immunity of a multicellular organism against internal and external threats to its integrity. In both situations, a coordination of many individuals (people and cells, respectively) responds to threats with suppressive or destructive systems. The origin of such group behavior and of group identity, however, must be understood from an evolutionary perspective. This chapter traces the early origins of biological identification and immune systems, first found in the prokaryotes, such as bacteria. A basal concept is developed and presented that bacteria can develop group identity and immunity and that this results from the action of colonization by genetic parasites (i.e., viruses). These parasites often employ "addiction modules" to attain stability. Addiction modules are matched sets of functions that require both antitoxic (protective/beneficial) responses to group members and toxic (destructive/harmful) responses to nonmembers. The role of such systems in the evolution of innate and adaptive immune systems of multicellular organisms is then traced. Similar concepts, however, also apply to the evolution of group identity in populations of higher animals, including social bonding and altruism. Here, the group recognition often depends on sensory cues, such as pheromones, as well as visual and audio sensory imprinting of behavior. With the evolution of the primates, sensory pheromone-based group membership was lost by genetic colonization and gene interruption (human DNA underwent invasion by genetic parasites). In humans, cognitive-based group identity replaced and supplemented these prior pheromone-based systems. Humans thus mostly attain group membership by developing states of "cognitive immunity" (i.e., belief.). Language and appearance are the main cognitive identifiers, and religion is the most prevalent form of social identity. The addictive and destructive potential of group identity, however, remains and can account for much conflict and threats between human groups.

42

Security can be defined as a form of group immunity that is protective of members but suppressive or destructive of internal and external threats to the social network. A security system thus seeks to maintain system integrity or status but allows growth or participation of new contributing members. This poses the dilemma that the destructive element should not also be self-destructive; it must recognize self. But what then is group identity and how can this be differentiated from threatening groups? How are we to identify a threat that has yet to happen? Can it be possible to predesignate entities or groups as foreign, or non-self, and therefore potentially harmful? Or, can we use a harmful event itself as a trigger for a group definition and subsequent defensive response? These questions can be posed to both societies and biological systems.

Biological systems must employ a security or immunity systems to ensure their survival. However, biological systems, unlike societies, are the products of evolution; they are ancient, spanning billions of years, and have survived numerous calamities. As the survivors of an evolutionary process, can they also instruct us concerning the best designs for attaining security for society? Can they suggest novel, optimized or more robust, autonomous, and local strategies for our own security? Biological systems are inherently robust, local, rapid, and adaptable systems. They are able to quickly marshal all the needed diverse and central resources but inherently reduce resource consumption when no longer needed. They are capable of searching for, finding, destroying, and sterilizing threats, both hidden and apparent. They are even able to respond to threats never before seen. However, they inherently limit self-destructive responses. Their self-limiting nature requires no instruction from a central system. They learn from past encounters with a threat and are consequently able to more rapidly and vigorously respond if a prior agent should reappear (via memory). The analogy of security to biological systems is therefore strong. Furthermore, biological systems have some characteristics that could clearly improve security systems if they could be incorporated. This chapter outlines the evolution of biological models of group identity and immunity and presents a thesis that links biological identity systems to security systems. In particular, security, immunity, and group identity are presented as highly related concepts, all of which depend on an ancient and enduring strategy that uses addiction modules. Human social identity, the source of many security concerns, has retained these basic biological strategies.

## Our Bodies as a Security System

The human body is able to respond to internal and external threats, including never before seen threats such as new cancers and new infectious agents. Although humans possess both the innate and adaptive immune systems (see glossary for definitions) found in most vertebrate animals, in addition, humans have an overlaying mental (cognitively mediated) behavioral capacity that allows even more rapid group responses to threats. However, our immune systems do not depend on a central authority, such as our brain, to initiate a response or secure the needed resources for a systemic response. Our immune systems do this automatically, against old or new threats, with no central instruction. Can such a biological capacity now inform us and help to better understand and anticipate the nature of threats faced by a modern society? Specifically, can the evolutionary nature of our biological security instruct us toward new models of security? To evaluate these questions, we need to understand how such sophisticated systems of self-identity and immunity evolved in the first place.

## Death of the Individual for the Good of the Whole

The main focus of this chapter is to outline the evolution of biological group identity and immunity with the goal of understanding human group identity systems. However, the evolution of cooperative group behavior poses the dilemma of how such selection is attained. Selection for an individual organism will often appear to be at odds with the selection for a group of organisms, especially groups where individual survival is not promoted. This same dilemma applies to the individual cells of a multicellular organism. Cell lineages (organs) that do not directly contribute to new offspring (germ, soma) have no future and are committed to death, so why should they contribute to good of the whole organism?

In the context of specific tissues, we can consider the differentiation and functioning of the human central nervous system (CNS) or immune system. Both depend heavily on programmed cell death (apoptosis), in which the vast majority of cells (undifferentiated neurons, bone marrow T cells) must die in the process of developing a functional tissue (CNS or immune system). How does such self-sacrificing cooperation evolve? Since cooperation is an inherent characteristic of group behavior, the question is really, how does group behavior or group identity evolve? This issue has often been considered from the perspective of how an organism can ensure the highest probable survival for its own and closely related genes. The models invoked have included ideas of kin selection, partner fidelity, and game theory in an effort to explain cooperation and overall group behavior. However, these models do not apply to understanding how tissues, such as the immune system, evolved. In general, the idea of group selection per se is not considered as a valid or experimentally supported concept. Thus, programmed cell death, which appears altruistic, presents an unsolved problem. In addition, most social structures of animal behavior also involve group behavior, which more often than not includes cooperation. Can the group behavior and cooperation of cells and organisms be explained by any common concepts?

In a recent book (Villarreal 2005), I developed and presented the concept that the accumulation and persistence of "genetic parasites" (i.e., viruslike parasites of the host genetic system, see glossary), which permanently alter the genetic identity of a host, has been an important creative force in the evolution of all life. In order for these genetic parasites to attain stable persistence in their host, they use a strategy called an addiction module, which is essentially a new identity system that provides both toxic (destructive/harmful) responses to nonmembers and antitoxic (protective/beneficial) responses to group members. It is precisely such identity systems that also endow their host with the capacity for group identity and the ability for cooperative or even altruistic self-destruction. The evolution of such group identification and destruction systems is the focus of this chapter.

## Persistence and Addiction: A Basal Concept

It will surprise most readers to realize that social theory applies to the most primitive life forms, such as bacteria, which contain systems capable of cooperation, group identity, and immunity. This makes sense, however, if we understand that it is of central importance for any organism to be able to identify itself and its mates or competitors. Bacteria have immunity systems that are both potentially destructive and protective. We now know that the most prevalent immunity systems of prokaryotes can be considered as examples of toxic/antitoxic (T/A) gene pairs (Engelberg-Kulka and Glaser 1999; Hayes 2003; Engelberg-Kulka et al. 2004; Lemos et al. 2005; Zielenkiewicz and Ceglowski 2005; see Table 4.1 for overview) of group identity systems in evolution. However, many of these systems have developed as a result of viruslike agents (persistent genetic parasites) found within the host genome that use T/A genes. The T/A genes ensure the stability of the parasite within the host by creating an "addiction state" in which the host (e.g., a bacteria) cannot lose the parasite and remain viable. Thus, T/A gene pairs (see box as restriction/modification [R/M] enzymes), (Table 4.1) are the most primitive form of so-called addiction modules but are critical to self and non-self identification and appear in various forms across all kingdoms of life, from bacteria to humans.

Addiction modules can also induce self-destruction. This concept seems illogical, especially for a single-celled organism, and an apparent violation of Darwinian principals with respect to survival of the individual. Yet, an illustration from the viral world shows the utility of these addiction modules for creating immunity (for references, see Villareal 2005). P1 is a common virus (phage) that typically persists in its host by expressing several addiction gene

## Development of Toxin/Antitoxin Systems

Some of the earliest experiments with pure cultures of bacteria revealed group immunity and group behavior. The process, known as lysogeny, was seen when two cultures of similar bacteria were grown together. One culture was able to produce a toxic virus (phage) that would kill (lyse) the second culture. The virus-producing culture was immune (via antitoxin immunity genes) to the phage due to phage immunity genes that had been incorporated into the bacterial DNA. Thus, this situation displayed both destructive and immune elements: two elements basic to the concept of a toxin/antitoxin (T/A) gene pair. Numerous other mixed-culture experiments identified additional agents that would similarly kill related but competing populations of bacteria while providing protection against self-destruction (Yarmolinsky 1995; Gelfand and Koonin 1997; Ichige and Kobayashi 2005). The consequence of such T/A modules were that groups (populations) of bacteria could recognize and kill other groups of bacteria. For the most part, these protein toxins (such as bacteriocins) also appear to have evolved from genetic parasites (prophage lysis genes) (Nakayama et al. 2000; Michel-Briand and Baysse 2002). However, in order for these toxin-producing bacteria to survive the very toxins they make, they needed to have genes that confer immunity to the toxin. For example, all prokaryotes use R/M enzymes as a molecular system of DNA immunity, able to recognize and destroy (restrict) foreign (unmodified) DNA. The potentially toxic restriction enzyme, however, is prevented from destroying the DNA of the immune host by the transient action of a DNA methylation (modification) enzyme. Since the restriction enzyme is stable but the modification enzyme is not and only acts during DNA replication, any new or foreign DNA will likely lack modification and be destroyed, but modified host DNA is protected.

sets. However, P1 colonized bacteria *(Escherichia coli)* will also self destruct if a second phage (i.e., T4) infects this host (Jensen and Gerdes 1995; Ramanculov and Young 2001). The disruption of the P1 T/A gene set is caused by expression of T4 genes and induces programmed cell death early in the infected cell program. The result of this is that any P1-colonized cell will quickly destroy itself if it becomes T4 infected, preventing T4 replication and transmission to the other P1-colonized E. coli (Xu et al. 2004). In this case, the P1-induced programmed cell death clearly resembles altruism. One cell in a related population is destroying itself to protect that population. Group identity and group immunity have both been attained and,

TABLE 4.1. Evolution of Group Identity Systems

| | Archebacteria | Cyanobacteria | Algae | Fungi | Nematodes | Bony fish | Land animals | Monkeys | Great apes | Humans |
|---|---|---|---|---|---|---|---|---|---|---|
| **Prokaryotes** | | | | | | | | | | |
| R/M | + | + | +/− | − | − | − | − | − | − | − |
| Prophage/lytic phage | + | + | − | − | − | − | − | − | − | − |
| Toxin/antitoxin | + | + | + | − | − | − | − | − | − | − |
| Bacteriocin/antibacteriocin | + | + | − | − | − | − | − | − | − | − |
| Chronic phage/immunity | + | − | − | − | − | − | − | − | − | − |
| Q/S pheromones | (+) (lactones) | + | − | + (peptide) | + | + (steroid) | + | + | − | − |
| **Eukaryotes** | | | | | | | | | | |
| ROS/antioxidant | · | · | + | + | + | + | + | + | + | + |
| RIP | · | · | · | + | − | − | − | − | − | − |
| RNAi | · | · | · | + | + | + | + | + | + | + |
| Apoptosis ID | · | · | · | · | + | + | + | + | + | + |
| −CNS (memory ID) | · | · | · | · | + | + | + | + | + | + |
| −Adaptive immunity (immune ID) | · | · | · | · | · | + | + | + | + | + |
| VNO receptor ID | · | · | · | · | · | + | + | + | − | − |
| Oxytocin/vasopressin | · | · | · | · | · | + | + | + | + | + |
| Secondary visual group ID | · | · | · | · | · | + | + | + | + | + |
| Red color vision (sex-linked) | · | · | · | · | · | · | · | + | + | + |
| Primary visual group ID | · | · | · | · | · | · | · | · | + | + |
| Audio learning (language) | · | · | · | · | · | · | · | · | · | + |
| Cognitive ID (beliefs) | · | · | · | · | · | · | · | · | · | + |

NOTE: CNS, central nervous system; Q/S, quorum-sensing; RIP, repeat-induced postmiotic; R/M, restriction/modification; RNAi, RNA interference; ROS, reactive oxygen species; VNO, vomeronasal organ. Dot (·) is not applicable, + is present, − is absent.

indeed, passed on to the *E. coli* host by a virus, giving the host a new genetic group identity in which individual cells are compelled to destroy themselves when that group identity is violated.

With this crucial although somewhat complex example, we see the overlay of several basic but inseparable issues that can now be emphasized. The stable persistence of a new virus provides a new genetic identity from a parasite-encoded T/A gene set that acts as an addiction module, conferring group immunity through self-destruction of the cell when its addiction module is disrupted. Thus, persistence, group identity, and group immunity are all fundamentally linked in a trinity. The concepts of kin selection or game theory need not be invoked as group identity, and individual programmed death (altruism) is a direct consequence of the resulting addiction. The suicide of the individual is an expected outcome of such group identity. With this trinity of linked concepts in hand, we can now evaluate the evolution of group immunity and identity in higher organisms.

## Adaptive Immune Systems of Worms

*Caenorhabditis elegans* (a nematode worm) is a basal higher organism (metazoan) of only about 1000 cells and is a highly studied model organism for the origin and development of a nervous system, as well as for neuromuscular-based motility and behavior. *Caenorhabditis elegans* has a highly efficient and adaptive immune system called the RNAi (RNA-interfering) system that can respond specifically to the sequences of any invasive genetic agent (virus). This "adaptive" immune system appears capable of protecting *C. elegans* from all known viral parasites (Barstead 2001; Bagasra and Prilliman 2004; Murakami et al. 2005; Schott et al. 2005; Wilkins et al. 2005) and is thus a major innovation in the evolution of cellular immunity. Ironically, the origin of this system appears to be viral in that it arose in response to endogenous retroviruses (distant relatives of HIV known as ERVs [endogenous retrovirus]) and other such genetic parasites that colonize and persist in all *C. elegans* genomes. In other words, genetic parasite colonization led to a silencing system, creating more complex identity systems to control the cell fates of neurons.

With the genetic colonization of *C. elegans* (ERV colonization), we see a general characteristic that has been maintained in the genomes of all metazoan animals (Ganko et al. 2001). Genetic parasites are often associated with group identity systems. ERVs and related genetic parasites are seen in all animal genomes, generally at substantial levels, and they can be more numerous than the genes of the organism. C. elegans' RNAi system appears to represent a basal RNA based immune system that led to the evolution of other RNA-based immune systems, as well as a cell-fate control systems.

Such systems are, in turn, basal components of both the nervous system and the immune system of all animals (see below).

## Apoptosis and a Nervous System

It seems ironic that the development of a nervous system required the cellular self-destruction process known as apoptosis. Yet this is precisely the case. A major innovation seen in *C. elegans* is the origin of central nervous tissue able to control whole-organism behavior, especially sex and group behavior. However, the development of a central nervous tissue and all other neurons depends on the process of apoptosis (cellular self-destruction) of some immature cells in order for cells to fully differentiate into terminal and functional neurons. At the same time, apoptosis itself is induced in connection to sensory- and growth-promoting signals for sensory nervous tissue. Thus we see the need for destructive (apoptosis) and beneficial (growth promotion) components working together to create a new, more complex identity system (in this case, a central nervous system) (Buss and Oppenheim 2004; Putcha and Johnson 2004; Chisholm and Jin 2005). It is interesting, then, that a main input that controls the fate and apoptosis of neurons is sensory receptors (G-proteins) on the cilia of sensory neurons (Zwaal et al. 1997), which detect pheromones, odors, and light. These sensory cilia are conserved in humans. Thus, environmental signals will program neurons to determine mating and feeding behavior (Peckol et al. 2001, Uchida et al. 2003). Similar receptors are also involved in detecting the density of other *C. elegans* (group detection) (Antebi 2005). Therefore, the invention of a terminal nervous system, and especially the associated development of environmental sensing units, has allowed the evolution of more complex group identities and behaviors.

Mechanistically, apoptosis itself, so crucial to the development of the sensory memory of the nervous system (and memory of the adaptive immune system), clearly has the elements of a T/A module (via disruption of mitochondria). Most cells express the genes needed for self-destruction (apoptosis) but are inhibited from apoptosis by the expression of matching inhibitors, and this system will self-destruct when the T/A dynamic is disrupted.

## Adaptive Immunity

The adaptive immune system is the most complex of all cell-based systems of group identity. Adaptive immunity can be highly informative concerning the basic principles for the design of a security system. Adaptive immunity identifies all the cells in the body as self and all others as non-self. It is then able to respond to essentially any foreign agent (old or new, harmful or not) that has not been already identified as self. It can then select for, and proliferate, cells that can bind and directly kill invasive cells or induce the production of large quantities of antibody molecules able to diffuse, bind to, and inactivate foreign antigens. The adaptive immune system can also be considered as a highly elaborate T/A system that is capable of a highly destructive but stable action, kept in check by the transient (time-dependent) education of group members. It directly depends on, and uses, cell growth and death (apoptosis) to shape the immune outcome and control itself. The adaptive immune system evolved with the fish (jawed vertebrates), after the evolution of the CNS in worms (lower metazoans). Like the nervous system, the adaptive immune system also uses large-scale self-destruction (apoptosis) to select for experienced-based development, proliferation, and memory of immune cells. However, unlike the CNS system, immune cells are genetically selected to be able to recognize and bind to antigens. Thus, they employ evolutionary principles of genetic variation, selection, and adaptation to ultimately select for the most effective immune responsive cells.

The adaptive immune system also uses apoptosis to control and limit the highly destructive nature of the immune response, and to limit resource consumption when these cells are no longer needed. It does this by the large-scale killing of self-reactive immune cells early during the development (education) of the immune system and also the killing of differentiated immune cells that are no longer antigen stimulated, resulting in only a small fraction of surviving immune cells. These cells recognize self but do not respond to self-antigens. The immune system is said to undergo education (tolerance) during this period of self-selection (see box for the imprinting of group identity). The priming, or initial learning, of the immune system is dependent on specialized cells (agents) that acquire the antigen and then specifically present the new antigen to immune cells. Certain conditions, such as trauma and cell death, as well as signals from the "first responders" or innate immune cells, will significantly stimulate the presentation of the antigen. This dependence on additional signals also helps to limit the potential of an overactive and self-destructive immune response.

However, like security systems, the immune system can also be subverted. In fact, it is rather curious that several viral families seem especially able to infect cells of the immune system. HIV, for example, uses the very receptors and the stimulation of the proliferative immune response to support viral infection and replication. In addition, sometimes the immune response itself creates the greatest pathology to the organism, as some agents are not very harmful, but a strong immune reaction can be harmful. For example, with hepatitis C viral disease the liver destruction is mainly due to a strong immune response and not destruction by the virus.

*Identity Transfer Is Temporally (Developmentally) Restricted*

A key question for understanding the persistence of group identity is under what conditions can such identity be transferred to *new* group members? In biological systems, group identity can be transferred to, or imprinted onto, new group members only during allowed periods. In the replication/modification system used by bacteria, only during DNA replication can the modification enzyme methylate the daughter cell DNA and protect it from degradation. This protective transfer occurs just prior to activation of the destructive part of the T/A module. In the adaptive immune system, this process is called the education of the immune system, in which self-identity is tolerated and protected from subsequent self-destruction. During this temporal window, the protective (unstable) component of an addiction module is first transferred, followed quickly by the transfer or activation of the destructive (stable) component of the addiction module. Both positive and negative selection with respect to identity can operate at this time. Both innate and adaptive immune systems tend to have this temporal characteristic. For example, apoptosis, innate complement-mediated immunity and T-cell and B-cell immunity are all potentially highly destructive systems that must be held in check by some form of self-protection that is acquired early on.

The immune system thus presents a clear and close analogy to defense. A traumatic event initiates the immune process and starts a strong immune reaction, and the immune system surveillance for known and new agents is continuous. While both these issues are of inherent interest for security, the immune system does have some additional features that could also be important for security systems. For one, it inherently operates by a local initiation process that can quickly engage a systemic response. This process is not predefined. Resources are taken as needed, not assigned. The response is inherently evolutionary and continually self-selects for a more efficient and diverse response. It uses addiction strategies, however, to limit the nature and duration of the response. The immune system does not tend to maintain a high level of activity when it is no longer needed. It is inherently self-limiting and does not depend on central command to scale it back. When the threat is gone, the system decays, but it retains a stable memory and thus subsequent response is rapid.

Although the adaptive immune system tells us a lot about designing a systemic method of defense, it does not really inform us much about the threats themselves or how biological threats might provide insights into security issues. Why do we face threats from various groups? Groups of people are not cells and are not directed by the types of biochemical interactions that control immune cells. Still, even human groups must evolve and follow some evolutionary principles. It would seem that some concepts that apply to biological pathogen groups, such as virulence, persistence, adaptation rates, and colonization, should also apply to these human groups. Human groups are under the control of social (and neurophysiological) principles that affect group behavior. What then can we say about evolutionary principles as they apply to the social group behavior of populations? How do these issues relate to the evolution and function of the biologically based group control?

## Basal Social Structures, Mating Behavior, and Mate Identification

Teleost (bony) fish represent a highly successful vertebrate that has both an adaptive immune system and clear group behavior. These fish are by far the most numerous vertebrates on the planet. Although most live in shoal populations (up to 30 billion in the case of Pacific herring), it is their reproductive and sexual behavior that is particularly fascinating. It is well known, for example, that individuals of some species (e.g., salmon) will essentially all commit suicide in order to swim back to their freshwater spawning grounds and procreate. Clearly such group behavior is difficult to explain if selection is simply limited to the success of the individual fish. For example, cheaters that fail to cooperate and choose nonlethal spawning grounds can easily be imagined and are predicted by theory. However, if reproductive membership and behavior are compelled by addictive (T/A-like) behavioral circuits of the CNS, we might be able to better understand the selective forces that can promote such apparently self-destructive outcomes.

We now know that reproductive behavior of jawed fish is also under pheromone control. Goldfish and zebrafish have olfactory sensory receptors similar to what was reported above with *C. elegans* (Dulka 1993; Cao et al. 1998; Zufall et al. 2005). Like *C. elegans,* fish also have ciliated sensory neurons (with G-protein receptors) that provide cues for mating behavior. The vomeronasal organ (VNO) is an externally exposed sensory neuronal tissue that has been identified as one of the important control elements of such behavior (Yoon et al. 2005). In some cases, fish can also form mating pairs. They also individually differentiate each other based on major histocompatibility complex (MHC) makeup (these antigens are an ID component of the adaptive immune system). In addition, it is clear that other sensory input, such as other soluble peptide and steroid molecules in the water, or the color and pattern of shoal members, provides cues concerning group membership (see Table 4.1, secondary visual group ID). It even appears that most jawed fish use a VNO-based system for reproduction, mating, and returning to spawning grounds. Thus the induced behavior is powerful enough to be suicidal.

This basic VNO circuit to control mating behavior via pheromone signaling has been highly conserved and maintained in all land animals (terrestrial tetrapods) (Zufall et al. 2005). Through mammalian studies we see that these systems might also be involved in the evolution of mammalian group identification. Lab mice *(Mus musculus)* are able to identify other individual mice that are either mating pairs or their own offspring. They will generally respond aggressively to nonmembers (mates, pups) introduced into their cage,s and this response is under VNO receptor control (Leypold et al. 2002; Freichel et al. 2004). It was long known that odorant-like (volatile) compounds in mouse urine were mediating much of this behavior, and it had been assumed that small molecule pheromones would be involved. Since most land animals (tetrapods) use urine marking to identify group membership and sexual behavior, it seemed likely that these processes should be conserved through evolutionary development. In mice, the VNO receptor uses the same MHC peptides as used by the adaptive immune system. Thus there is also a direct overlap between how the immune system and the nervous system attain group identity.

### Primates/Humans and Group Identity

A plausible scenario is that all terrestrial tetrapods use VNO systems to identify group (family, sex partner) membership, supplemented by visual and other cues. The needed markers of individuality are often secreted, sometimes in the urine but sometimes by specific scent glands. Thus, most mammals scent-mark their habitat for social purposes, and some of the marking components are derived directly from components of the immune system. In addition to such pheromone-based social bonding, it is also likely that other sensory inputs (via visual-audio sensory cilia G-proteins) are similarly used for the purpose of sex and social group membership identification. For example, both visual color patterns and songs are often used in specific bird and fish species for flock, shoal, and mating-pair identification. It seems likely that such inputs would similarly need to permanently modify CNS circuits to stably affect behavior. Such a scenario appears precisely to characterize the CNS response observed in adolescent male song birds during audio learning and imprinting. Can this scenario apply to human group behavior?

Because the VNO system is exceedingly well conserved in the evolution of most vertebrates, it is a major surprise to learn that the great apes, including humans, have inactivated essentially all of their VNO genes (Giorgi et al. 2000; Kouros-Mehr et al. 2001; Liman and Innan 2003; Zhang and Webb 2003). If such a system has been so basic for group membership, as implied above, how can the great apes attain complex group social dynamics without this system? Notably, great apes and humans do not depend on scent marking (as do New World monkeys, which have retained the VNO system) to identity themselves or groups. The divergence of these two primate lineages is on the scale of 5 million years. Coincident with this divergence are other major developments. First, the great apes evolved a third red light–sensitive photoreceptor, found on the X chromosome (see Table 4.1). Second, great apes were increasingly invaded by a specific family of genetic parasites in the form of endogenous retroviruses known as HERVKs (human endogenous retrovirus) (as well as other genetic parasites, LINES [long interspersed nuclear elements], SINES [short interspersed nuclear elements], and alu's), which may have incapacitated the VNO system during their disruptive colonization. This invasion was particularly widespread in humans, who have substantially more HERVK elements than chimpanzees, including eight types that are specific to humans (Barbulescu et al. 1999; Sverdlov 2000). Third, the brain expanded substantially in the great apes, especially the neocortex. Interestingly, the human neocortex also expresses numerous human-specific endogenous retroviral sequences not found in the other great apes (Karlsson et al. 2001; Nakamura et al. 2003; Frank et al. 2005; Perron et al. 2005). Moreover, the human neocortex is enlarged, is much more invasive (via connections) of basal brain and spinal cord structures, and has more specialization for visual and audio input than the neocortex of other mammals (Striedter 2005). Since the neocortex is considered the seat of expanded human consciousness and intelligence, including social intelligence, is there any way we might understand all these dizzying associated changes? And if so, can it offer any explanation for human group identification? The following scenario proposes such a linkage.

The genetic invasion that incapacitated the primate VNO system would seem to represent a highly destructive event. Since social and sexual group behavior is essential for all species, but is in fact enhanced in great apes, clearly their VNO loss would select for the development of other systems able to compensate for and replace the lost social identification previously provided by the VNO system. In the case of the primates, it seems clear that visual, cognitive-based behavior provided this solution. A major process by which primates identify members is via visual recognition. The acquisition by great apes of the third photoreceptor (G-protein) associated with color (red) vision, and curiously maintained on the X sex chromosome (Smallwood et al. 2002; Collin and Trezise 2004; Kawamura and Kubotera 2004), is thus of great interest (relative to the other terrestrial tetrapods, which maintain only two photo receptors that are not sex associated). The presence of this receptor in primates suggests the importance of visual recognition, especially in reproductive bonding. Although visual recognition was clearly an already well-developed capacity in vertebrates, it does not appear that most vertebrates depend much on visual processes for social bonding (birds are a notable exception). For example, gestures and facial expressions in particular are prominent in chimpanzee social recognition (Parr et al. 2000; Parr 2003). If this sensory input was then used as a main conduit for social identification, clearly visual reception and associated vision-based CNS systems must have undergone compensatory adaptation. Brain-imaging studies in fact support the idea that chimpanzees, like humans, do indeed have specialized brain regions that are able to respond to facial and gestural patterns. Thus we can propose that the great apes adopted mainly visual sensory input as a primary system to replace the prior pheromone-based input for social identification. This visual dependence would also be consistent with the need to develop and expand the necessary brain structures so that much more complex visual sensory information and pattern recognition could be used for a social communication. The expansion of the primate brain and neocortex was thus under positive selection for this essential group identification purpose. The resulting chimpanzee social structures are therefore able to include much more complex sensory cues and patterns, rather then being limited to MHC peptides or other pheromones. An enhanced cognitive capacity was thus essential for this complex visual pattern recognition, which must also be learned and retained (imprinted) in a stable, group-defining way.

## Primate Social Structures, Cognitive Imprinting

Continuing along these lines, humans underwent not only another expansion of their brain size, but also a disproportionate enlargement of their neocortex. The human neocortex also shows a greater invasive innervation of other brain and spinal cord structures. Additionally, humans acquired audio learning and complex vocalization. In a sense, we can argue that human evolution extended the visual cognitive capacity that had evolved in all the great apes (for social identification) and added a vocal cognitive (learning/speaking) system that could also be used for group identification. Interestingly, to a large degree the human brain has also segregated these audio and visual functions into the left and right hemispheres, another distinction from our primate relatives. As with chimpanzees, functional magnetic resonance imaging (fMRI) studies also support the idea that humans have specialized CNS systems dedicated to the detection of facial expressions and patterns (Okada et al. 2003; Loffler et al. 2005). The white part of the human eye, for example, is thought to have been an adaptation that aids in the recognition of gazing. Standard facial expressions (fear, joy, anger, etc.) are inherently recognized by all cultures, and the classification of this process has recently become a digital science for the animation industry. Thus, an almost hard-wired and universal human facial recognition capacity exists, including not only a relatively standard range of emotion states, but also the very rapid recognition of racial face types (Kim et al. 2006). Appearance (facial expressions, racial appearance) would then have been the main sensory input used during the early evolution of human group identity. Additionally, human audio learning is distinct from that of other audio-learning animals in that it is recursive; that is, meaning is variable and dependent on the order of otherwise identical words. Thus, learning language requires much more complex pattern recognition than learning songs, for example, and may relate to why distinct brain structures are involved in singing and speaking the same words. Humans are much more adaptable in terms of complex audio pattern recognition, including the abstract patterns of meaning, than any other animal.

Thus, it could be argued that human group or social recognition systems are basically cognitive and not biochemical in nature as they are in all other tetrapods. Early human social structures depended primarily on audio and visual inputs to identify social group members and used enhanced CNS systems to induce and imprint stable cognitive and behavioral identity modules that act as addiction modules. Like all the other systems of group identity and immunity we have so far considered, this imprinting also occurs during a specific time window of development (or experience, see box).

In terms of audio learning and group identity, language has a strong and clear association with group membership. Language acquisition during a critical developmental window of childhood clearly results in altered brain structures and enhanced cognitive capacity. A "normal" human must learn a language to develop a fully human cognitive capacity. Feral children that never learn language become incapable of subsequently acquiring language or of developing deeper cognitive functions that are considered inherent human characteristics. In fact, it appears that different languages (e.g., English vs. Chinese) may lead to different types of brain alterations (Tham et al. 2005). If we propose that language is basically an audio- and cognitive-based system of group identification, we might also be better able to explain the very large diversity of languages associated with early human cultures and the seemingly unending group conflicts also associated with language.

## Cognitive Identity and Immunity

The evolution of cognitive-based human group recognition and membership raises a whole other set of deep questions as we seek to understand the character and implications of this state. Like all identity and immune systems, in order for a cognitive-based group membership system to be stably maintained, it must recognize and exclude nonmembers. It should also incorporate the typical beneficial and harmful elements (T/A addiction module) for stability and identity. What does this mean in the case of cognition?

Cognition is the mental process by which sensory information is processed, via thought and recognition, and stored or evaluated for action. In order for cognition to be used as a basis of group identity or immunity, it must be open to sensory input and mental processing from recognized sources (members), but resistant to sensory input or mental response from nonmembers. Since audio learning is specific to the human primates (Sherwood et al. 2003), let us focus on audio group identity first. Assuming language is a main audio identity system, it would need to be recognized, acquired, and elicit the appropriate mental processing (thought/meaning) in members but not nonmembers. These are exactly the characteristics of someone who has learned and recognizes a specific language. During a crucial developmental period in childhood, the CNS has been "colonized" and structurally modified (imprinted) by the process of learning the language. The resulting cognitive state is open to sensory communication (language input) from members, but not from nonmembers. In addition, it is stable and not easily lost or displaced. However, as alluded to above, there is perhaps an unintended cognitive consequence of having learned to hear and speak a language. The resulting restructuring of the neocortex needed for such complex pattern recognition has also created a neocortex with a more sophisticated inherent pattern recognition capacity. Thus, the cognitive capacity itself has been significantly enhanced by language that also presents additional opportunities for cognitive-based group identification (a positive feedback loop of cognition).

This cognitive social identity (addiction) scenario, it is proposed, can account for the forces that led to early human social structures, which were mainly extended families or tribal in nature. With the evolution of common language, it was possible for larger social structures and identities to evolve. Acquiring these cognitive identities has also expanded the capacity of the brain to perform complex pattern recognition, including allowing the ability to read a written language.

Since a written language is read via visual input, but uses an audio-based complex language skill as a mental system of pattern recognition, learning to read cross-references two distinct cognitive structures in the human brain. Like the acquisition of complex audio learning, learning to read leads to a restructured brain that appears to have also significantly enhanced pattern recognition. This, I will argue, also resulted in an unintended but enhanced platform for cognitive identity systems to develop.

One necessary outcome of the relationship of language to group identity is that the cognitive meaning of language must be set and stable. Stable memories (meaning) must be stored in a way that is not easily lost or displaced. This has a very interesting and deep implication. Once meaning is given to an audio pattern (word), that meaning must be resistant and not easily be lost. Meaning, we could say, has a "belief status," if belief is defined as a stable assignment of meaning. The view that belief-based (stable) sensory input is stored in a distinct way and place from nonbelief (novel) inputs in the human brain is supported by fMRI studies of humans (Goel and Dolan 2003; Grezes et al. 2004).

At this point it is worth reconsidering the definition of "belief" and its evolutionary origins. The assertion made above is that humans differ from other animals because of their highly enhanced capacity to use cognitive systems for group identity. The origin of a belief (or stable) state for sensory information is inherent and necessary in its use as a system of cognitive identification. Belief can also be defined as a state of cognitive immunity, in which meaning is stably held and resists other sensory inputs or identifiers. Those of us who are involved in education often feel that new knowledge should be accepted by students if it can simply be communicated clearly and the transmitted information shown to be verifiable in the real world (such as evolution). Recent studies, however, make it clear that many people are often prone to beliefs of many things that are clearly contrary to verifiable information. The implication of the above argument is that belief states are more biologically fundamental than generally realized. Any newly presented knowledge will be resisted if it conflicts with stably stored belief information. Thus education (acquisition of new knowledge) need not result in the displacement of belief states, since these are inherently stable identifiers. Related concepts, cognitive dissonance, belief perseverance, and social identity are thus the psychological consequences of more biologically basal cognitive immunity. We can now consider the role of belief as a system of cognitive immunity that provides the broadest group identity.

## Religion, Cognitive Immunity, and Group Identity

Of all beliefs, religious beliefs appear to be among the most stable, although many other belief systems are also enduring. Once fully accepted (imprinted), such beliefs can apparently resist essentially all conflicting sources of sensory input. Religious belief structures were especially stabilized by the invention of written language and also appear to be mainly transmitted by writing. Notably, an early definition of the word "literate" meant being able to read holy letters and words of the Bible. A belief is a purely cognitive group identifier needing no biological or other sensory characteristics to be imprinted. It can include any tribal, racial, language, and cultural groups, including gangs of otherwise identical populations. Religious beliefs can be considered as an especially highly addictive and stable example. Group membership is attained by cognitive acceptance. One must "simply" believe to belong.

## Pheromone-Based Pair-Bonding and Toxin/Antitoxin Systems

The ability of some animals to form stable social pair bonds has been much studied as it appears to relate to human social interactions and social identity. One well-studied system involved in this is the vomeronasal organ (VNO), a neuronal tissue in the nasal epithelia that recognizes pheromones associated with pair bonding. The VNO neurons have also been shown to be involved in the social bonding between mating pairs of various mammals (voles). Unlike birds, the majority of mammals do not form stable, lifelong pair-bonds with their mates. Some voles, however, are clear exceptions to this. Pair-bonding of the prairie vole is known to occur during the initial extended mating of individual virgin females to males (Young et al. 2005). Following this imprinting experience, the bonded male will refuse to mate with other females and will respond aggressively to them. The behavioral mechanism of this aggressive response is not known. Pair-bonded females will respond similarly, although the central nervous system (CNS) mechanism involved appears to differ between the sexes. It was established, by severing the VNO nerve process to the brain, that the VNO neurons were directly involved in this social bonding (Meek et al. 1994; Curtis et al. 2001). In addition, neurophysiological analysis has indicated the oxytocin (Bales et al. 2004; Smeltzer et al. 2006) induction during mating is directly involved and that this induces a permanent and elevated expression of the vasopressin receptor in regions of the male brain. Interestingly, these regions correspond to the same regions involved in drug (opiate) addiction (Eisenstein 2004). Surprisingly, the resulting pair-bonded state between mates could also be artificially induced in species (mice) that do not normally pair-bond. When an artificial recombinant virus expressing the vasopressin receptor was introduced into this region of the male brain while being associated with the female (imprinted), measurable pair-bonding resulted (Landgraf et al. 2003). This result has very interesting implications for the evolution of behavior-based group identification. A virus was able to affect social group behavior. The mountain vole is a very similar species to the prairie vole but does not undergo mated pair-bonding. The biggest genetic difference between these vole species is seen on their sex chromosomes, which harbor distinct sets of parasitic sequences. In terms of the vasopressin receptor control, an invasion by a genetic parasite (satellite DNA) into the promoter region of this gene appears to have resulted in a different regulatory response to VNO signaling (Hammock et al. 2005; Pennisi 2005). Here then, too, is evidence that genetic parasites might have been involved in the events that led to the different CNS social responses in these two species of voles. The possibility also seems strong that behaviorally addictive CNS circuits are involved in stable social identification. However, in this case, the applicable T/A module would be behavioral, mediated by matched aggressive and associative behavior. This has major implications to human behaviors, particularly those that are highly stable and associated with strong group identification. However, as we will see below, humans do not use VNO systems to attain social bonding and identification.

One of the strongest associative social bondings in humans is the romantic love between mates. In the prairie vole, addictive CNS circuits were stably altered (via receptor transcription) during mating and depended on VNO sensory input for mate identification. The result was a socially bonded pair, involved the same circuits as drug addiction. I consider this bonding equivalent to an immune state since it is beneficial to identified members, but aggressive and harmful to nonmembers (nonmates). However, it has been clear since these vole studies were first reported that human pair-bonding could not be the same. Clearly without a VNO system, human mate recognition cannot depend on the same pheromones. However, recent functional magnetic resonance studies do support the idea that humans who are in intense romantic love with their mates have distinct CNS responses when shown pictures of their mates, but not other pictures (Bartels and Zeki 2000, 2004; Najib et al. 2004; Aron et al. 2005, Fisher et al. 2005). In addition, areas of human brain activation, as with the voles, are also associated with drug addiction. This observation is consistent with the proposal that humans use visual sensory inputs, along with T/A-like addiction modules to initiate pair-bonding states. Also consistent with this T/A idea is the intense and immediate psychological pain that can result from a broken love bond. As religions also often depend on group feelings of love, it seems likely they represent cognitive T/A modules.

Religious memberships, like immune systems, clearly have a T/A character in that they are potentially harmful to nonmembers and supportive of members. The harmful component can sometimes be virulent (often lethal) to nonmembers. The protection of members can be compassionate and at times altruistic, involving mortal defense sacrifice of individual members for the good of the whole. In this, religions can sometimes promote an apoptotic-like state in which individual members will readily engage in potentially mortal conflict and even destroy themselves for the maintenance of group identity (belief). This constitutes the largest and most expansive cognitive immune/identity system in existence, encompassing populations of billions. The tendency toward religious belief thus appears inherent to human cognitive systems, as they were clearly held by all known human cultures, even, apparently, Paleolithic cultures. The militant Islamists currently threatening world security are tightly defined by their religious beliefs, differentiated from even other forms of Islam. The resemblance to a virulent genetic colonizer seems strong and seems to represent a virulent version of group identity that is attacking nonmembers.

## General Lessons for Security

The analogy of security to biological immune and group identity systems appears to be highly applicable. The basic evolutionary principles used in biological systems are surprisingly well maintained and can be seen to apply across the broadest spectrum of biology, from bacteria and their viruses to human populations. Group identity and immunity are intermingled concepts, and both require destructive and protective components. In fact, a direct chain of reasoning and evidence links the biological group identification systems to the very origin of the virulent versions of religious group identification. It is belief systems, essentially the human form of T/A-type addiction modules, that currently threaten world security. The Islamo-fascist belief appears to represent a virulent variant of group identity that attacks all nonmembers. It has a strongly toxic and self-sacrificing (apoptotic) character that provides a very stable (cognitively addictive) group identity. It clearly demonstrates the absolute power of belief over the survival of the individual that is likewise inherent in many addiction modules. It has a metastatic and invasive character with implications for global social security.

What strategies can biology suggest to confront this situation? Can virulent group members undergo a process selection for avirulence or of deaddiction? For example, can pharmacology or psychoactive compounds affect such states of hostility or absence of empathy? If such states do indeed have biological foundations, such biological interventions would also seem plausible. However, our understanding of these cognitive process is so poor, applications from current technology are highly questionable. It is clear that educational imprinting, which develops rational and ethical thinking modes, can also counter virulent group (belief) identity. However, educational imprinting likely has developmental restrictions that work best when applied to children, adolescents, or young males. Can an educational deprogramming approach work on a mature immune mind? We do not know, since stable (and thus irreversible) alterations in brain structure might be involved in creating this group immunity.

The immune response also suggests some strategies. The current threat situation resembles that posed by a latent but virulent pathogen as it hides, waiting to attack its host and promoting its own replication by hijacking a larger group identity. For example, the radical Takfiri Muslim belief of al-Qaeda seeks to overtake the identity of more moderate Muslims. Once established, such viruslike latent infections are difficult for even the most advanced immune system to eliminate. However, sterilizing immune responses, including autoimmunity, can often be initiated by trauma, and even a latent infection will sometimes resolve following an eruption. Can cognitive immunity also respond this way to trauma? This seems possible. For example, Muslims (including Sunnis) generally suffer most casualties following nonspecific terrorist attacks, such as the wedding bombings in Jordan on Nov. 11, 2005, that killed 57 and wounded 100, organized by Abu Musals al-Zarqawri from Iraq. This "autotrauma" compelled King Abdullah of Jordan to dissociate from the radical Takfiri al-Qaeda belief and declare war on such beliefs. It was in fact the subsequent Jordanian intelligence that was crucial to the successful killing of al-Zarqawi on June 8, 2006. However, moderate/rational Muslim groups often hold some level of (auto)identity with the radical groups. But as this example shows, once moderates have been attacked by the radical group, it is likely that members of this moderate population are transiently susceptible to stimulation by the trauma to dissociate their identity from that of the radical groups (a kind of autoimmunity). Such events should more effectively be used as an adjuvant to stimulate an antiradical moderate Muslim reaction. So far, such moderate populations have not tended to react strongly to such attacks, indicating a high tolerance to self, or innocent destruction. It also seems that radical Arab public relations has characterized such destruction as permissible and within the bounds of group identity and will sometimes even convince susceptible believers that it was foreign directed. It is unlikely, however, that those that have personally experienced such trauma can feel this way. Given the ability of politics and the advertisement/entertainment industry to affect and manipulate popular beliefs, it seems this capacity has not been well applied here. The absolute cruelty of these groups (beheadings, assassinating children and relatives, etc.) is not well exploited by public relations systems to dissociate group identity.

One thing seems clear, strong religious belief is a stable state of cognitive immunity, and altering this belief identity is not likely to result from strictly rational discourse, as it stems from deep biological foundations inherent to human cognition. When virulent, such beliefs are clearly viruslike and demand a societywide response, as does a virulent epidemic. To address this, the relevant issue is not Christian versus Muslim identity. Any belief group can potentially generate virulent versions (e.g., Nazis, Ku Klux Klan, Kmer Rouge) relative to less virulent versions of nationalism and racial pride, although susceptibilities to virulence vary. Rather, organized and sponsored transmission of such virulent identity to new generations should no longer be tolerated anywhere in the world. Such situations can no longer be considered as simply a local problem. Like the outbreak of Avian influenza in remote or poorly developed regions, a worldwide response is needed. Virulent group identity threatens to infect the entire civilized world. Models based on biological immunity and control could help design a coherent but decentralized worldwide response.

## ACKNOWLEDGMENTS

I wish to acknowledge Raphael Sagarin for providing the initial motivation to write this chapter plus his strong editorial assistance that compelled me to write a more focused chapter.

## GLOSSARY

Biological scientists require precise terminology in order to accurately communicate observations and ideas. However, this same terminology is often not understood by the general audience and may be labeled as "jargon." Yet the use of more generalized wording will often result in inaccurate or misleading statements. The terms below provide definitions that should be understood by the lay audience. For more precise definitions see Villarreal (2005).

*Adaptive immunity.* An immune system, like that of mammals, that can adapt to, recognize, and respond to new foreign agents and infections.

*Apoptosis.* A cellular system of mitochondrial-mediated self-destruction associated with the development of many tissues, such as the central nervous system and the adaptive immune system.

*Bacteriocins.* Toxins, made by bacteria, that are generally poisonous to other related bacteria.

*Cognitive ID.* The human capacity to establish stable cognitive states regarding the acceptance of information. A state also known as belief.

*Cyanobacteria.* Photosynthetic bacteria (blue-green) found in the oceans.

*Group identity.* The capacity of a group in individual cells or individuals of one species to differentiate themselves from other, usually closely related groups.

*Group immunity.* The ability of a group of cells in one organism of a group of individuals within one species to recognize and respond to non-group members. The immune response is generally defensive and destructive.

*Genetic parasites.* A genetic agent (i.e., virus) that is parasitic to the information system (copying and translation) of its host cell.

*Holins.* Toxins made by viruses (phage) of bacteria that are used to lyse the infected bacteria and release the newly replicated virus.

*Innate immunity.* An immune system, like that of simple animals, that can pre-recognize foreign agents or infections but does not adapt to new agents

*Oxytocin.* A hormone associated with maternal and other behavior in animals.

*Prophage.* A virus of bacteria that becomes silent and part of the bacterial chromosome but provides immunity against similar virus.

*Q/S pheromones.* Quorum-sensing pheromones. Small molecules released by bacteria used to sense the presence of other related bacteria.

*Reactive oxygen species.* Bacteria with a set of chemical responses derived from oxygen that are used as toxins against foreign agents. Their toxicity is controlled by antioxidants.

*Repeat-induced postmiotic system.* A DNA defense system found in fungi that recognizes and inactivates copied DNA (genetic parasites).

*R/M (restriction/modification) system.* A coupled system of two enzymes found in bacteria that recognizes and degrades foreign DNA.

*RNAi.* An RNA-based defense system (interference) found in many eukaryotes that recognizes and degrades forgein RNA.

*Vasopressin.* A hormone associated with maternal and other behavior in animals.

*Visual group ID.* The ability of humans and great apes to visually recognize group members.

*VNO receptor.* A protein present on specific neurons of the olfactory mucosa, the vomeronasal organ, that functions as a receptor for pheromones associated with sexual behavior in animals.

## REFERENCES

Antebi, A. 2005. The prepared mind of the worm. *Cell Metabolism* 1 (3): 157–158.

Aron, A., H. Fisher, D. J. Mashek, G. Strong, H. Li, and L. L. Brown. 2005. Reward, motivation, and emotion systems associated with early-stage intense romantic love. *Journal of Neurophysiology* 94 (1): 327–337.

Bagasra, O., and K. R. Prilliman. 2004. RNA interference: The molecular immune system. *Journal of Molecular Histology* 35 (6): 545–553.

Bales, K. L., A. J. Kim, A. D. Lewis-Reese, and C. S. Carter. 2004. Both oxytocin and vasopressin may influence alloparental behavior in male prairie voles. *Hormones and Behavior* 45 (5): 354–361.

Barbulescu, M., G. Turner, M. I. Seaman, A. S. Deinard, K. K. Kidd, and J. Lenz. 1999. Many human endogenous retrovirus K (HERV-K) proviruses are unique to humans. *Current Biology* 9 (16): 861–868.

Bartels, A., and S. Zeki. 2000. The neural basis of romantic love. *Neuroreport* 11 (17): 3829–3834.

Bartels, A., and S. Zeki. 2004. The neural correlates of maternal and romantic love. *Neuroimage* 21 (3): 1155–1166.

Barstead, R. 2001. Genome-wide RNAi. *Current Opinions in Chemistry and Biology* 5 (1): 63–66.

Buss, R. R., and R. W. Oppenheim. 2004. Role of programmed cell death in normal neuronal development and function. *Anatomical Science International* 79 (4): 191–197.

Cao, Y., B. C. Oh, and L. Stryer. 1998. Cloning and localization of two multigene receptor families in goldfish olfactory epithelium. *Proceedings of the National Academy of Science U S A* 95 (20): 11987–11992.

Chisholm, A. D., and Y. Jin. 2005. Neuronal differentiation in *C. elegans*. *Current Opinions in Cell Biology* 17 (6): 682–689.

Collin, S. P., and A. E. Trezise. 2004. The origins of colour vision in vertebrates. *Clinical and Experimental Optometry* 87 (4–5): 217–223.

Curtis, J. T., Y. Liu, and Z. Wang. 2001. Lesions of the vomeronasal organ disrupt mating-induced pair bonding in female prairie voles *(Microtus ochrogaster)*. *Brain Research* 901 (1–2): 167–174.

Dulka, J. G. 1993. Sex pheromone systems in goldfish: Comparisons to vomeronasal systems in tetrapods. *Brain Behavior and Evolution* 42 (4–5): 265–280.

Eisenstein, M. 2004. Is it love . . . or addiction? *Laboratory Animals* (NY) 33(3): 10–11.

Engelberg-Kulka, H., and G. Glaser. 1999. Addiction modules and programmed cell death and antideath in bacterial cultures. *Annual Review of Microbiology* 53: 43–70.

Engelberg-Kulka, H., M. Reches, B. Sat, S. Amitai, and R. Hazan. 2004. Bacterial programmed cell death systems as targets for antibiotics. *Trends in Microbiology* 12 (2): 66–71.

Fisher, H., A. Aron, and L. L. Brown. 2005. Romantic love: An fMRI study of a neural mechanism for mate choice. *Journal of Comparative Neurology* 493 (1): 58–62.

Frank, O., M. Giehl, C. Zheng, R. Hehlmann, C. Leib-Mösch, and W. Seifarth. 2005. Human endogenous retrovirus expression profiles in samples from brains of patients with schizophrenia and bipolar disorders. *Journal of Virology* 79 (17): 10890–108901.

Freichel, M., R. Vennekens, J. Olausson, M. Hoffmann, C. Müller, S. Stolz, J. Scheunemann, P. Weissgerber, and V. Flockerzi. 2004. Functional role of TRPC proteins in vivo: Lessons from TRPC-deficient mouse models. *Biochemistry and Biophysics Research Communications* 322 (4): 1352–1358.

Ganko, E. W., K. T. Fielman, and J. F. McDonald. 2001. Evolutionary history of Cer elements and their impact on the *C. elegans* genome. *Genome Research* 11 (12): 2066–2074.

Gelfand, M. S., and E. V. Koonin 1997. Avoidance of palindromic words in bacterial and archaeal genomes: A close connection with restriction enzymes. *Nucleic Acids Research* 25 (12): 2430–2439.

Giorgi, D., C. Friedman, B. J. Trask, and S. Rouquier. 2000. Characterization of nonfunctional V1R-like pheromone receptor sequences in human. *Genome Research* 10 (12): 1979–1985.

Goel, V., and R. J. Dolan 2003. Explaining modulation of reasoning by belief. *Cognition* 87 (1): B11–B22.

Grezes, J., C. D. Frith, and R. E. Passingham. 2004. Inferring false beliefs from the actions of oneself and others: An fMRI study. *Neuroimage* 21 (2): 744–750.

Hammock, E. A., M. M. Limm, H. P. Nair, and L. J. Young. 2005. Association of vasopressin 1a receptor levels with a regulatory microsatellite and behavior. *Genes Brain and Behavior* 4 (5): 289–301.

Hayes, F. 2003. Toxins-antitoxins: Plasmid maintenance, programmed cell death, and cell cycle arrest. *Science* 301 (5639): 1496–1499.

Ichige, A., and I. Kobayashi. 2005. Stability of EcoRI restriction-modification enzymes in vivo differentiates the EcoRI restriction-modification system from other postsegregational cell killing systems. *Journal of Bacteriology* 187 (19): 6612–6621.

Jensen, R. B., and K. Gerdes 1995. Programmed cell death in bacteria: Proteic plasmid stabilization systems. *Molecular Microbiology* 17 (2): 205–210.

Karlsson, H., S. Bachmann, J. Schröder, J. McArthur, E. F. Torrey, and R. H. Yolken. 2001. Retroviral RNA identified in the cerebrospinal fluids and brains of individuals with schizophrenia. *Proceedings of the National Academy of Science U S A* 98(8): 4634–4639.

Kawamura, S., and N. Kubotera 2004. Ancestral loss of short wave-sensitive cone visual pigment in lorisiform prosimians, contrasting with its strict conservation in other prosimians. *Journal of Molcular Evolution* 58 (3): 314–321.

Kim, J. S., H. W. Yoon, B. S. Kim, S. S. Jeun, S. L. Jung, B. Y. Choe. 2006. Racial distinction of the unknown facial identity recognition mechanism by event-related fMRI. *Neuroscience Letters* 397 (3): 279–284.

Kouros-Mehr, H., S. Pintchovski, J. Melnyk, Y. J. Chen, C. Friedman, B. Trask, and H. Shizuya. 2001. Identification of non-functional human VNO receptor genes provides evidence for vestigiality of the human VNO. *Chemical Senses* 26 (9): 1167–1174.

Landgraf, R., E. Frank, J. Aldag, I. D. Neumann, C. A. Sharer, X. Ren, E. F. Terwilliger, M. Niwa, A. Wigger, and L. J. Young. 2003. Viral vector-mediated gene transfer of the vole V1a vasopressin receptor in the rat septum: Improved social discrimination and active social behaviour. *European Journal of Neuroscience* 18 (2): 403–411.

Lemos, J. A., T. A. Brown Jr., J. Abranches, and R. A. Burne. 2005. Characteristics of *Streptococcus mutans* strains lacking the MazEF and RelBE toxin-antitoxin modules. *FEMS Microbiology Letters* 253 (2): 251–257.

Leypold, B. G., C. R. Yu, T. Leinders-Zufall, M. M. Kim, F. Zufall, and R. Axel. 2002. Altered sexual and social behaviors in trp2 mutant mice. *Proceedings of the National Academy of Sciences U S A* 99 (9): 6376–6381.

Liman, E. R., and H. Innan 2003. Relaxed selective pressure on an essential component of pheromone transduction in primate evolution. *Proceedings of the National Academy of Sciences U S A* 100 (6): 3328–3332.

Loffler, G., G. Yourganov, F. Wilkinson, and H. R. Wilson. 2005. fMRI evidence for the neural representation of faces. *Nature Neuroscience* 8 (10): 1386–1390.

Meek, L. R., T. M. Lee, E. A. Rogers, and R. G. Hernandez. 1994. Effect of vomeronasal organ removal on behavioral estrus and mating latency in female meadow voles *(Microtus pennsylvanicus)*. *Biology of Reproduction* 51 (3): 400–404.

Michel-Briand, Y., and C. Baysse 2002. The pyocins of *Pseudomonas aeruginosa*. *Biochimie* 84 (5–6): 499–510.

Murakami, M., T. Ota, S. Nukuzuma, and T. Takegami. 2005. Inhibitory effect of RNAi on Japanese encephalitis virus replication in vitro and in vivo. *Microbiology and Immunology* 49 (12): 1047–1056.

Najib, A., J. P. Lorberbaum, S. Kose, D. E. Bohning, and M. S. George. 2004. Regional brain activity in women grieving a romantic relationship breakup. *American Journal of Psychiatry* 161 (12): 2245–2256.

Nakamura, A., Y. Okazaki, J. Sugimoto, T. Oda, and Y. Jinno. 2003. Human endogenous retroviruses with transcriptional potential in the brain. *Journal of Human Genetics* 48 (11): 575–581.

Nakayama, K., K. Takashima, H. Ishihara, T. Shinomiya, M. Kageyama, S. Kanaya, M. Ohnishi, T. Murata, H. Mori, and T. Hayashi. 2000. The R-type pyocin of *Pseudomonas aeruginosa* is related to P2 phage, and the F-type is related to lambda phage. *Molecular Microbiology* 38 (2): 213–231.

Okada, T., S. Tanaka, T. Nakai, S. Nishizawa, T. Inui, Y. Yonekura, J. Konishi, and N. Sadato. 2003. Facial recognition reactivates the primary visual cortex: A functional magnetic resonance imaging study in humans. *Neuroscience Letters* 350 (1): 21–24.

Parr, L. A. 2003. The discrimination of faces and their emotional content by chimpanzees *(Pan troglodytes)*. *Annals of the New York Academy of Science* 1000: 56–78.

Parr, L. A., J. T. Winslow, W. D. Hopkins, and F. B. de Waal. 2000. Recognizing facial cues: Individual discrimination by chimpanzees *(Pan troglodytes)* and rhesus monkeys *(Macaca mulatta)*. *Journal of Comparative Psychology* 114 (1): 47–60.

Peckol, E. L., E. R. Troemel, and C. I. Bargmann. 2001. Sensory experience and sensory activity regulate chemosensory receptor gene expression in *Caenorhabditis elegans*. *Proceedings of the National Academy of Science U S A* 98 (20): 11032–11038.

Pennisi, E. 2005. Genetics. In voles, a little extra DNA makes for faithful mates. *Science* 308 (5728): 1533.

Perron, H., F. Lazarini, K. Ruprecht, C. Péchoux-Longin, D. Seilhean, V. Sazdovitch, A. Créange, N. Battail-Poirot, G. Sibaï, L. Santoro, M. Jolivet, J. L. Darlix, P. Rieckmann, T. Arzberger, J. J. Hauw, and H. Lassmann. 2005. Human endogenous retrovirus (HERV)-W ENV and GAG proteins: Physiological expression in human brain and pathophysiological modulation in multiple sclerosis lesions. *Journal of Neurovirology* 11 (1): 23–33.

Putcha, G. V., and E. M. Johnson Jr. 2004. Men are but worms: Neuronal cell death in *C. elegans* and vertebrates. *Cell Death and Differentiation* 11 (1): 38–48.

Ramanculov, E., and R. Young 2001. Functional analysis of the phage T4 holin in a lambda context. *Molecular Genetics and Genomics* 265 (2): 345–353.

Schott, D. H., D. K. Cureton, S. P. Whelan, and C. P., Hunter. 2005. An antiviral role for the RNA interference machinery in *Caenorhabditis elegans*. *Proceedings of the National Academy of Science U S A* 102 (51): 18420–18424.

Sherwood, C. C., D. C. Broadfield, R. L. Holloway, P. J. Gannon, and P. R. Hof. 2003. Variability of Broca's area homologue in African great apes: Implications for language evolution. *Anatomical Record A: Discoveries in Molecular and Cellular Evolutionary Biology* 271 (2): 276–285.

Smallwood, P. M., Y. Wang, and J. Nathans 2002. Role of a locus control region in the mutually exclusive expression of human red and green cone pigment genes. *Proceedings of the National Academy of Science U S A* 99 (2): 1008–1011.

Smeltzer, M. D., J. T. Curtis, B. J. Aragona, and Z. Wang. 2006. Dopamine, oxytocin, and vasopressin receptor binding in the medial prefrontal cortex of monogamous and promiscuous voles. *Neuroscience Letters* 394 (2): 146–151.

Striedter, G. F. 2005. Principles of brain evolution. Sunderland, MA: Sinauer.

Sverdlov, E. D. 2000. Retroviruses and primate evolution. *Bioessays* 22 (2): 161–171.

Tham, W. W., S. J. R. Liow, J. C. Rajapakse, T. C. Leong, S. E. S. Ng, W. E. H. Lim, and L. G. Ho. 2005. Phonological processing in Chinese-English bilingual biscriptals: An fMRI study. *Neuroimage* 28 (3): 579–587.

Uchida, O., H. Nakano, M. Koga, and Y. Ohshima. 2003. The *C. elegans che*-1 gene encodes a zinc finger transcription factor required for specification of the ASE chemosensory neurons. *Development* 130 (7): 1215–1224.

Villarreal, L. P. 2005. *Viruses and the evolution of life.* Washington, DC: ASM Press.

Wilkins, C., R. Dishongh, S. C. Moore, M. A. Whitt, M. Chow, and K. Machaca. 2005. RNA interference is an antiviral defence mechanism in *Caenorhabditis elegans. Nature* 436 (7053): 1044–1047.

Xu, M., D. K. Struck, J. Deaton, I.-N. Wang, and R. Young. 2004. A signal-arrest-release sequence mediates export and control of the phage P1 endolysin. *Proceedings of the National Academy of Science U S A* 101 (17): 6415–6420.

Yarmolinsky, M. B. 1995. Programmed cell death in bacterial populations. *Science* 267 (5199): 836–837.

Yoon, H., L. W. Enquist, and C. Dulac. 2005. Olfactory inputs to hypothalamic neurons controlling reproduction and fertility. *Cell* 123 (4): 669–682.

Young, L. J., A. Z. Murphy Young, and E. A. Hammock. 2005. Anatomy and neurochemistry of the pair bond. *Journal of Comparative Neurology* 493 (1): 51–57.

Zhang, J., and D. M. Webb. 2003. Evolutionary deterioration of the vomeronasal pheromone transduction pathway in catarrhine primates. *Proceedings of the National Academy of Science U S A* 100 (14): 8337–8341.

Zielenkiewicz, U., and P. Ceglowski. 2005. The toxin-antitoxin system of the streptococcal plasmid pSM19035. *Journal of Bacteriology* 187 (17): 6094–6105.

Zufall, F., K. Ukhanov, P. Lucas, T. Leinders-Zufall. 2005. Neurobiology of TRPC2: From gene to behavior. *Pflugers Archives* 451 (1): 61–71.

Zwaal, R. R., J. E. Mendel, P. W. Sternberg, and R. H. Plasterk. 1997. Two neuronal G proteins are involved in chemosensation of the *Caenorhabditis elegans* Dauer-inducing pheromone. *Genetics* 145 (3): 715–727.

Part Three

# SECURITY TODAY

Chapter 5

# CORPORATIONS AND BUREAUCRACIES UNDER A BIOLOGICAL LENS

ELIZABETH M. PRESCOTT

Security threats have evolved to more closely mimic natural pressures that confront organisms and ecosystems. In nature, complex challenges require dynamic solutions that benefit from being tested in the evolutionary marketplace. These natural defense strategies offer insight into national and international security strategies for confronting nontraditional security challenges. Decision makers must respond to these evolving situations while under selective pressure from the political ecosystem. Better understanding of natural defense strategies will inform development of the most effective responses to complex threats, allowing for better integration of lessons learned from nature.

## Nontraditional Challenges to Global Security

The cold war approach to security entails countering symmetrical threats by actors that were large, identifiable, and prioritized, making them amenable to being compartmentalized for management separately and sequentially.[1] The world now faces security challenges that are complex, networked, numerous, and equally devastating, making prioritization of response much more difficult. Many threats have the ability to undermine the viability of a nation, or the world, making it unwise to delay attention to any one of them. Further, these threats demand interconnected responses that prioritize allocation of resources without loosing ground in countering the plethora of other challenges. Organizationally, this requires speed, integration, and less hierarchy of decision making, allowing semiautonomous components to respond rapidly to evolving challenges.

As the security environment evolves, strategies to counter challenges must also adapt to counter new challenges. The concept of "Darwinian security"—the use of evolutionary lessons to inform national and international security strategies—aims to aid this process. Drawing lessons from nature and the survival of species in harsh and evolving environments offers lessons for creating security strategies more adapted to meeting current challenges.

The concept of security incorporates a breadth of issues with the overarching aim of preventing harm or disruption to a population. The changing nature of global interconnectedness has created new vulnerabilities to individuals and nations. Nontraditional security threats such as health, environment, and food security, can undermine the well-being of a population, challenging global stability and security.[2] These security threats have unique attributes that require countering with nontraditional security strategies and coordinating many semiautonomous components. Responses to challenges of this nature require participation by diverse actors including civil society, companies, and individuals from diverse segments of society. Engagement by the private sector and general citizenry is often critical to a success. For example, in devising strategies for countering the emergence of a human strain of pandemic influenza, a majority of the assets needed to assure the health and safety of the U.S. population—hospitals, transport networks, communication systems—are in the hands of the private sector. Additionally, most large societal disruptions require citizens to make critical decisions based on timely information. If government recommendations are not followed or feasible—such as the evacuation strategy for New Orleans before Hurricane Katrina—the necessary governmental response changes. The diversity of essentially semiautonomous actors in a dynamic and globalized world creates challenges to crafting effective strategies.

These networks of diverse actors afford the opportunity to use different tools than traditional defense and deterrence strategies to influencing outcomes. Harnessing the collective energy of these semiautonomous components into a productive response to a security threat is not easy. Nonstate actors have differing motivations for addressing security vulnerabilities, complicating efforts to create incentive strategies. Each of these actors must accept the premise of the threat and the agreed upon strategy for response. Coordination between these stakeholders as well as the central authority requires synchronization, lending a powerful role to systems for information sharing and messengers who convey that information. These factors must be employed by governments to engage the semiautonomous segments of society to effect the appropriate response to nontraditional security challenges.

Many of these security challenges require taking action in the face of high levels of uncertainty. The multiplicity of semiautonomous actors in nontraditional security challenges—those that perpetrate and those that

## *Response to Global Security Challenges*

An example of a political response to nontraditional security challenges can be seen in the global response to the health challenge of H5N1 avian influenza. The appearance of this virus in Hong Kong in 1997 went largely unnoticed as a global security issue, but aggressive measures initiated against the virus allowed only temporary containment. The scientific community, keenly aware of the threat posed by a pandemic strain of influenza, continued to track the virus as it migrated around the world, awakening individual governments to the threat posed by this animal disease that had real and potential human health implications. Individual nations were forced to confront this global challenge with little assistance from the rest of the world. Most countries on the front line of the battle were ill prepared to deal with the challenge financially or politically. The scientific community sharpened their discourse in an effort to garner attention to the issues, but governments and the private sector remained unconvinced of the real threat.

In the wake of a calamitous hurricane season and a few pivotal policy moments in the summer of 2005, the world awoke to the microbial situation. Most nations responded by focusing on domestic preparations for the emergence of a human pandemic strain of influenza. Efforts to respond by governments or the international community were hampered by uncertainty in timing, scale, and expected manifestation of the threat. Meanwhile, the virus continued to propagate and was presented with ever-expanding opportunities to mutate into the form that is most feared. The private sector, recognizing the vulnerability of commerce to the impacts of a pandemic in a globalized world, began to make continuity plans to assure continued operations in the event of a global disruption. Collective anxiety resulting from the specter of high human health costs was magnified further by predictions of a global economic recession that might befall in the event of a moderate human pandemic.

The global response to avian influenza highlights some of the complications of dealing with an emerging security challenge. The tools commonly used to address global security problems have proven cumbersome and inadequate for devising a strategy to counter this threat. Governments initially unwilling to offer financial assistance eventually exceeded expectations at a pledging conference in Beijing in January 2006, but by then the virus had significantly spread so it was unlikely that available tools would have been enough to achieve eradication.

The heterogeneous social and cultural landscape has enabled the virus to thrive often in the regions least capable of responding to the challenge. Intergovernmental coordination between diverse agencies proved to be problematic for many countries. Weaknesses in communication between ministries of health and ministries of agriculture in countries hit with human cases of $H_5N_1$ were slow to be resolved. The former, tasked with assuring human health, were often driven by different motivations than the later, which were often geared more toward promotion of domestic agriculture interests.

As this nontraditional security threat continues to play out on the global stage, many of the challenges to dealing with a highly dynamic situation and coordinating a response can be seen. Much can be learned from biological models about how to deal with this viral challenge. A Darwinian response to the challenge of avian influenza would allow a more task-oriented response, allowing resources to be allocated to the location and function that are most critical to mitigating the threat at that point in the evolution of the challenge. Rather then every country allocating significant resources to domestic medical preparation in the event of a future pandemic, funds should be invested in prevention such as paying peasant farmers not to hide their sick chickens from authorities. Innovative solutions based on the most cutting edge science should be employed across national lines and facilitated in countries unable to domestically support such activities. Connecting decision makers, health and animal experts, and the private sector within nations and across nations in a dynamic response network is the best chance for efficiently addressing this challenge and could help prevent a human health catastrophe.

respond—creates more variables when assessing a challenge and initiating a response. As many of these actors are not traditionally considered part of the security community, expectations for behavior may not be fully formed. Also, the pace at which security challenges unfold requires swift and decisive action, forcing decisions to be made with less information than would otherwise be desired.

Interestingly, nontraditional security challenges can mimic threats that persistently confront organisms in nature. Organisms exist in complex ecosystems with many interconnected semiautonomous actors. An individual strives to prolong its life and propagate in hostile environments while balancing the cost of developing and utilizing security measures. Extant organisms are descendents of ancestors that achieved a sufficient balance between security and other needs. This balance results in security strategies that have already been tested in the "evolutionary marketplace" for their ability to aid in the prosperity of a specific individual organism. These strategies may give insight into balancing the costs and benefits of national and international security strategies that address similar challenges such as determining the acceptable level of risk from biological terrorism, knowing that eliminating the risk completely will be extremely costly to society.

This chapter explores types of uncertainty and evolutionary strategies devised to accommodate that uncertainly. The idea of a political ecosystem will be explored to highlight similarities with natural ecosystems, as well as pressures that prevent the emergence of dynamic and adaptable systems. Additionally, examples of evolutionary strategies being employed in security context will be discussed.

## Different Types of Uncertainty

Life does not afford perfect knowledge of future challenges, and organisms must adapt to compensate for this uncertainty. Strategies to deal with various types of uncertainty have been employed throughout evolution. These solutions to natural challenges can inform the processes policy makers use to approach uncertainty surrounding national and international security challenges.

### Predictable Challenge with Timing Uncertain: Biological Cascade

When confronted with an event that has been envisaged as a threat but with uncertain timing, coordination of the complex network of semiautonomous national security actors and resources is critical. Natural phenomena such as droughts, disease, floods, and other extreme events are unpredictable in timing but can be expected to occur over time. For example, a severe weather system was seen as a threat to New Orleans before Hurricane Katrina, but the timing was uncertain, and, as a result, the ability to focus political will among competing threats was greatly diminished. In nature, one strategy for dealing with this problem is to create cascades of events to yield a predetermined outcome.

In a cascade, specific molecules are activated to perform a predetermined function that works in conjunction with other actors.[3] A cascade often commences with a single molecular actor that goes on to magnify the impact through the activation of additional molecular actors. Finely tuned stimulators and repressors induce the appropriate action for each molecular actor, synchronizing the response and allowing for concerted action at many levels of the system. Some molecular actors have flexibility in the degree of activation; in other words, the final state does not have to be on or off but rather can be attenuated to achieve the desired outcome. Further, the triggers and responses have been predetermined through a process of molecular trial and error. This enables a broad array of predetermined responses allowing greater flexibility in countering uncertainty in timing.

A cascade allows an organism to quickly react to subtle stimuli by mounting a response to a previously identified trigger. One well-characterized cascade sequence is that of blood coagulation.[4] Coagulation is critical when an organism is confronted with an unexpected injury that results in bleeding. To respond to this threat, humans have evolved an elaborate system to allow a multiplied and synchronized response to stabilize and begin the healing process. First, after an injury, blood vessels are constricted, slowing blood flow to the sight of injury. This is followed by activation of many molecular actors that flood the scene, performing predetermined functions. Some of these actors are able to adapt to meet the specific needs of the injury by changing shape to precisely fill the wound. Subsequently, a repair plug is stabilized with a protective mesh, or clot, that prevents the plug from dislodging. Finally, after tissue repair has been completed, the clot must be dissolved to restore normal blood functioning. If these steps are not properly performed or controlled, the individual's life can be put in jeopardy, making the cascade of events critical to the viability of the organism.

Lessons from the biological cascade model can be applied to security challenges that are foreseeable but with uncertain timing, such as natural disasters or terrorist attacks. In nature, trial and error determine the appropriate response allowing the most successful strategy to evolve over time. This is unrealistic in a national or international security environment, where experimentation with a high likelihood of failure with high human costs would not be politically palatable. However, a cascade model can be useful when the appropriate response to the challenge is known and the relevant actors can be identified. A trigger to begin the cascade needs to be established with additional actors understanding their appropriate role in response. Completing the cascade requires following the predetermined activities to completion, making reliability a possible point of critical weakness. From these parameters it is possible to devise the most effective plan for a specific event through simulation cascades and coordinate the response. While not all security vulnerabilities can be envisaged, those that are—flooding, earthquakes, and other disasters—could be better planned for by understanding how nature responds to challenges that present similar uncertainties.

## Challenge or Response Uncertain: Heterogeneity and Redundancy

In international security, it is not always possible to predict a challenge or determine the appropriate response in advance. Faced with similar challenges, nature has devised strategies to respond to unforeseen threats by generating heterogeneity and redundancy in critical systems.

Redundancy in critical systems is necessary to assure needed functions can be performed in the event of changing circumstance. In many life forms, alternate strategies for meeting vital functions—where failure can undermine viability—have evolved. Energy generation is an example where duplicative systems have emerged allowing adaptation to changes in resource ability. In the single-cell yeast *Saccharomyces cerevisiae,* the preferred source of energy is a plentiful type of sugar called glucose.[5] Glucose is generally prevalent in the environment, and the metabolism of glucose is an efficient source of energy for the cell. However, when glucose is not present and alternate sugars are available, the organism engages a different and less efficient set of genes to produce the needed energy.

Equally, heterogeneity allows for more options available to respond to unpredictable selective pressure. Diversity of options is beneficial in two ways. On one hand, different options increase the likelihood that there will be an available solution to resolve the problem when it arises. On the other hand, subtle differences also make it less likely that all solutions will be unsuccessful. For example, heterogeneity in genetic material differentiates individual organisms in the species. If faced with an unpredicted challenge, diversity in genetic characteristics makes it more likely at least some individuals will be equipped to counter the new challenge. At the same time, genetic heterogeneity also lowers the probability that all individuals in the species will be eliminated in the face of a challenge, presenting the species with a better chance of surviving to the next generation.

In policy however, duplicative or heterogeneous strategies can be difficult to defend, as they appear to be an inefficient allocation of resources. For example, there is constant pressure to reduce the cost associated with provision of health care. Resources that are left idle—surplus beds, machines, and hospital facilities—are viewed as wasteful and are often eliminated. From a systems perspective, however, extra capacity can be critical in the event unexpected failures make it difficult to access previously available facilities. In absence of redundancy in capabilities, a critical medical device being off-line could have significant negative health impacts.

Achieving heterogeneity in policy can also be difficult. Governments often make single solution policy decisions that require significant time and resources to implement. Policymakers craft solutions to satisfy interested groups based on available information at a point in time, oversimplifying dynamic challenges with imperfect information. Having embarked upon a policy course, it is often difficult to alter the path or employ competing strategies. This creates a homogenous policy response, lowering diversity in approach and reinforcing critical weaknesses.

Comparatively, the private sector has an advantage in maintaining a more heterogeneous environment, as multiple self-interested actors make decisions based on changing information. A profit-making company looks to differentiate itself from competitors through strategic decisions based on available information. The size of an investment is often adjusted for confidence in the predicted outcome. If a company wants to enter into an industry, a small investment can be made to test the waters. If it is successful, the endeavor can be expanded to increase exposure. Similar analysis by multiple actors can cause market decisions to become synchronized, creating vulnerabilities. Overall, however, the market allows experimentation in high-risk endeavors while still allocating resources based on changing information about the economic environment. This creates the opportunity to tinker with current strategies, allowing for more heterogeneity than a single, large actor can achieve.

Nature employs many different solutions simultaneously, with selection occurring after the strategy has been expressed at the level of the individual organism. As a result, most solutions will not be the most successful. On the surface it appears that this trial-and-error process invests resources less efficiently than a single, considered decision. However, in complex systems it is difficult to predict the full impact of small attenuations. A solution that optimizes one variable could inadvertently result in externalities in other parts of an interconnected system. In nature this trade-off can be seen in genetic traits such as sickle cell anemia. This small genetic mutation confers protection against malaria when a human receives a single genetic contribution from a parent but is lethal when both parents contribute the mutation.[6] The mild disease caused by a single genetic dose is not optimal, but the heterogeneity created presents opportunities for an individual to survive other life threatening challenges.

Although it is unrealistic and undesirable to strive for completely heterogeneous allocation of resources or redundancy in all security systems, it is critical to remember that complex, interconnected systems have unique characteristics that need to be looked at in context of the entire security environment. Policy and security interventions can have unpredictable externalities. Understanding processes by which nature accommodates for these unknown outcomes can be useful in addressing these challenges.

## Scale Uncertain: Surge Capacity

In nature, organisms also must confront uncertainty in the scale of a challenge making it critical to determine the appropriate level of resource allocation without creating undue negative side effects. This balance requires an accurate assessment of the real threat and the time frame in which the response needs to be mounted, recognizing that a disproportionate response can sometimes do more harm as failure to respond.

In nature, a prime example of scale uncertainty can be seen in the immune system. Upon activation, the immune system employs aggressive tactics in response to a challenge. This requires a delicate balance to avoid unnecessary harm to the organism from the powerful immunological tools employed. Many molecular actors are involved with the detection of pathogen and the subsequent response to the threat. Of these, cytokines play a critical role as messengers between various immune system cells, incorporating local feedback about the perceived nature of the threat.[7] These signals are very sensitive, allowing for real-time detection of a microbial challenge before the pathogen has become too numerous to counter with available immunological tools. At times, however, these molecules can exaggerate the nature of the threat, resulting in negative health impacts.[8] A "cytokine storm," or an over reaction by the immune system to the real threat, is thought to be involved with the acute immune response to microbes such as SARS and H5N1 avian influenza.[9] When confronted with a new microbial challenge, the immune system aggressively attacks the microbe, ultimately compromising the organism. Although a weak initial immune response can allow a pathogen to get a firm hold in the host, an overly aggressive initial response can result in death from the immune response before the pathogen has been able to cause serious illness.

There are parallels to confronting the appropriate scale of response in security challenges. It is not possible to eliminate all security risks, creating the need to assess, prioritize, and respond appropriately to recognized challenges. Many available tools for confronting security threats carry the risk of adverse consequences if applied inappropriately. An exaggerated political response can have negative implications. For example, the U.S. political response to the challenge of September 11 demonstrates the complexity of crafting policy remedies for perceived security weaknesses. After September 11, 2001, policy makers put forth aggressive legislative proposals in an effort to address perceived vulnerabilities. Some of these regulations, such as the U.S. Patriot Act of 2001 and the Bioterrorism Preparedness and Response Act of 2002, resulted in extensive alterations to a complex regulatory system. Changes to the regulatory framework for dealing with high-risk individuals and activities also carried negative externalities for legitimate international visitors and activities.[10] Prior to the event, security weaknesses were highlighted but were not given appropriate attention. The reforms that followed were swift but without extensive consideration for the impact on other government priorities. Although the balance will always be delicate, greater understanding of nature's strategies for managing scale of response can inform political efforts to avoid scaling failures. For example, political systems that allow for policy feedback from many different sources could enable inclusion of what might otherwise be unforeseen externalities resulting from policy decisions. Engaging a broad array of stakeholders in discussions could facilitate unexpected feedback about potential implications of policy reforms.

## Political Ecosystem

Much like natural ecosystems, the political environment evolves in response to local political pressures. Tip O'Neill Jr., former Speaker of the House, claims that "all politics is local"; this captures the significant influence that proximal environment plays in political outcomes.[11] Ideally, a democratic political ecosystem rewards efforts to optimize the allocation of resources to achieve a beneficial outcome for society and punishes those that squander resources. The political process should confer rewards to those who accurately predict current and future needs and craft efficient strategies to meet those needs.

In practice, the political ecosystem is far more complex. Selective pressure for determining fitness is based on the opinion of those being governed, with expectations playing a critical role. Expectations are diverse and based on disparate criteria, but stepping too far out of equilibrium with popular expectations can negatively impact political leaders. Preferences and priorities influence interpretation of the appropriate action and allocation of resources. Depending on which vantage point one is observing from—the politician who desires to be reelected or the constituent who cares about remaining employed or the activist with a deeper knowledge of the relevance of specific issue areas—the fitness of a political decision will be viewed differently. Further, in the political ecosystem, the future is heavily discounted. This distorts efforts to achieve allocation of resources to maximize long-term beneficial outcomes.

Systematic rigidities also impact political ecosystems. In democratic societies political leaders must allocate resources in a way that is predictable and accountable. This is often achieved by setting expectation for the allocation of resources well in advance of the need for the actual expenditure. Adhering to previous promises builds trust but also creates obstacles to employing new solutions in a changing environment. As such, bureaucratic systems designed to assure accountability create impediments to allocating resources to counter unpredictable challenges. There are glaring exceptions to this, such as the U.S. funding for the military activities in Iraq or funding after Hurricane Katrina, but those represent a small portion of unpredicted challenges and also require significant political attention to accomplish.

## Resource Allocation

Many nontraditional security challenges involve high levels of uncertainty, creating problems for resource allocation in bureaucratic organizations. Predictable funding requires a deliberative process that is rigid and time-consuming. Once the priorities have been identified, pressure exists to meet those obligations, despite recognition of a changed environment.

For example, the United States begins determining the federal budget several years in advance of expenditure. This time lag is critical for negotiating priorities but presents little ability to response to unpredicted events. While this system can be effective at meeting the needs of less dynamic expenditures such as infrastructure or education, it is less effective at allocation for events with high levels of uncertainty. As a case in point, few would have predicted that the SARS virus was going to emerge, requiring extensive expenditures by governments and the international public health community. In order to meet this unfunded mandate, monies were diverted from other public health programs, leaving insufficient funds to meet other program goals. Neglecting these other public health functions could create further vulnerabilities to future health challenges.

In practice, public health professionals know emerging health threats require dynamic allocation of funds; however, the budget process does not enable allocation of funds in anticipation of unexpected health events. A similar failure of budgetary alignment can be seen with unpredicted security challenges. In the aftermath of the terrorist attacks in 2001, frantic political efforts shuttled funds through the allocation process to address acute vulnerabilities. Extraordinary political will was required despite the existence of a unique climate of consensus about the degree of risk. It is unrealistic to think that all security threats will meet the same level of compelling circumstances.

Further, the absence of a visible threat is often viewed as the absence of a threat. If a defense strategy is successful, a threat is not allowed to manifest. With many nontraditional security challenges, determining success requires relying on the absence of an event occurring that was uncertain to occur in the first place. For example, a successful public health program will lower the incidence of the targeted disease. In subsequent years, the resources allocated to maintain the program look large when compared with the burden of disease. In the zero-sum game of resource allocation, this makes the program a prime target for decreased funds. This approach may seem rational in the short term but overlooks the fundamental premise of defense systems that require maintenance in the absence of a observable need. Therefore, defense systems for threats that do not have visible and positive feedback are often victims of their own success.

This paradox creates a tension between the need to maintain apparently underutilized systems and the desire to allocate funds to meet what are seen as the current short-term needs. Critically, the cost saved by eliminating successful programs does not always compare favorably with the cost needed to reinstate the program at a later date. A case in point is the global public health defense against the smallpox virus. This disease has historically had significant human health impacts but was declared eradicated everywhere on the planet in 1970s.[12] Children ceased being vaccinated against the virus as part of the normal childhood vaccination process. Following September 11 and the anthrax attacks on Capitol Hill, the specter of bioterrorism rekindled attention on this virus, which was again perceived to be a significant human health threat. Unfortunately, after letting the vaccine defense system wane, it has not proved easy to raise the global immunity to smallpox, consuming tremendous resources and political attention.

Biological models can offer insight into dealing with decisions surrounding allocation of resources for defense systems. For example, the immune system must allocate resources carefully in the face of complex and uncertain events. When a threat is detected, a surge in resources is allocated to confront the perceive challenge. Once the system has neutralized the immediate threat, the system scales down to assume a basal level of coverage against that threat, keeping low levels of sensitive molecular detectors in circulation. While the threat is not present, it allows for swift scaling up if the threat were to subsequently be detected. The ability to craft defense systems that require minimum expenditure to maintain but are able to quickly surge would allow more efficient resources allocation against unpredicted challenges.

## Solutions to a Changing Security Environment

The changing nature of security needs is forcing the security ecosystems to evolve to become more networked, integrated, and adaptable. Independent initiatives such as the Princeton Project on National Security are working to devise a long-term national security strategy focusing on addressing systematic vulnerabilities rather than responding to specific threats. In the United States, the military has been progressive in networking systems to achieve real-time integration of intelligence and operational activity.[13] However, decades of technological integration into operational capacity have created a gap between military and civilian national security capabilities. This divide is likely to be at least partially responsible for the growing list of objectives—from humanitarian missions to border security—being reassigned from civilian to military operations, suggests the need to develop similar capabilities in civilian organizations.[14]

Reform efforts addressing civilian and diplomatic security capabilities in the United States appear to be incorporating characteristics observed from natural security strategies. On January 18, 2006, United States Secretary of State Condoleezza Rice announced "transformational diplomacy," which aims "to build and sustain democratic, well-governed states that will respond to the needs of their people—and conduct themselves responsibly in the international system."[15] To meet these objectives, the initiative incorporates evolutionary strategies to achieve more nimble and integrated systems, for example, increasing capacity for real-time diplomatic adaptation by placing political officers with military units in the battlefield for more accurate information gathering. In contrast to a hierarchical decision-making process directed from Washington, this could enable timely feedback critical to an effective response. Cooperation and coordination among diverse security actors is recognized as critical to success of a unified response to complex challenges. Much as signaling pathways need to be synchronized, national security professionals need to understand their counterparts in allied military, political, or intelligence organizations. Career development will need to be more adaptive to internalize the changing security needs. Integrating education over a career with counterparts outside of the primary agency will break down function-specific barriers in personnel and staffing. The ability to reallocate human capital to effectively manage resources creates a more dynamic system to respond to unpredicted circumstances. These changes recognize that modern security challenges are more episodic than traditional single-engagement conflicts, mimicking the evolutionary battle for survival in a hostile and changing world.

However, any successful effort to reform must effectively address the changing security environment while also maneuvering through the political ecosystem. In this context, this proposal has many hurdles to overcome. Personnel in large bureaucratic organizations often benefit from maintaining the status quo in a system they have successfully maneuvered. Ultimately, implementation of elements of transformational diplomacy could be an experiment in applying evolutionary strategies to security and diplomacy.

## Conclusion

The changing nature of security threats present challenges in crafting effective security strategies. These complex and dynamic nontraditional threats require new tools to establish and effect the appropriate response. Defense and response strategies that have survived in the evolutionary marketplace offer useful insight to dealing with interconnected events that involved many semiautonomous actors with high levels of uncertainty. Not all evolutionary strategies will be useful for all security challenges, but taking lessons form these systems and the political ecosystems in which they must function will inform the national and international security strategies for the new century.

NOTES

1. Leon Fuerth, "Strategic Myopia," *The National Interest,* Spring 2006, Issue 83.

2. For more information on nontraditional security threats please see Joseph S. Nye, *Paradox of American Power: Why the World's Only Superpower Can't Go It Alone* (New York: Oxford University Press, 2002).

3. Entry for "cascade sequence" in *Oxford Dictionary of Biochemistry and Molecular Biology* (New York: Oxford University Press, 1997), 96.

4. Review articles: H. M. H. Spronk, J. W. Govers-Riemslag, H. ten Cate, "The Blood Coagulation System as a Molecular Machine," *BioEssays* 25, no. 12 (2003), 1220–1228; Earl W. Davie "A Brief Historical Review of the Waterfall/Cascade of Blood Coagulation" *Journal of Biological Chemistry* 278, no. 51 (2003), 50819–50832.

5. Juana M. Gancedo, "Yeast Carbon Catabolite Repression," *Microbiology and Molecular Biology Reviews* 62, no. 2 (1998), 334–361.

6. More information on sickle cell anemia can be found at www.ncbi. nlm.nih.gov/books/bv.fcgi?call=bv.View..ShowSection&rid=gnd.section.98 (last accessed June 2007) or www.pbs.org/wgbh/evolution/library/01/2/ L_012_02.html (last accessed June 2007).

7. More information on cytokines can be found in Angus W. Thomson and Michael T. Lotze, *The Cytokine Handbook* (London: Academic Press, 2003).

8. Sang Heui Seo, E. Hoffmann, and R. G. Webster, "Lethal H5N1 Influenza Viruses Escape Host Anti-viral Cytokine Responses," *Nature Medicine* 8 (2002), 950–954, published online August 26, 2002; Chung Y. Cheung, Leo L. M. Poon, Iris H. Y. Ng, Winsie Luk, Sin-Fun Sia, Mavis H. S. Wu, Kwok-Hung Chan, Kwok-Yung Yuen, Siamon Gordon, Yi Guan, and Joseph S. M. Peiris, "Cytokine Responses in Severe Acute Respiratory Syndrome Coronavirus-Infected Macrophages In Vitro: Possible Relevance to Pathogenesis," *Journal of Virology* 79, no. 12 (2005), 7819–7826.

9. Michael T. Osterholm, "Preparing for the Next Pandemic," *New England Journal of Medicine* 352 (2005), 1839–1842.

10. Committee on Research Standards and Practices to Prevent the Destructive Application of Biotechnology, National Research Council, *Biotechnology Research in an Age of Terrorism* (Washington, DC: National Academies Press, 2004), available at www.nap.edu/catalog/10827.html (last accessed June 2007).

11. Tip O'Neill and Gary Hymel, *All Politics Is Local, and Other Rules of the Game* (Holbrook, MA: Adams Media Corporation, reprint edition, 1995).

12. Jonathan B. Tucker, *Scourge: The Once and Future Threat of Smallpox* (New York: Atlantic Monthly Press, 2oo1).

13. Leon Fuerth, "Strategic Myopia," *The National Interest*, Spring 2oo6, Issue 83.

14. Ibid.

15. www.state.gov/r/pa/prs/ps/2oo6/59339.htm (last accessed June 2oo7).

Chapter 6

# SELECTION, SECURITY, AND EVOLUTIONARY INTERNATIONAL RELATIONS

GREGORY P. DIETL

Alfred North Whitehead (1925, 179) once remarked that "change is inherent in the very nature of things." Despite the undoubted truth of this statement, many political scientists treat economic entities, such as states, as static and unproblematic units that move across the international stage, not unlike chess pieces (Cederman 1997). Change (new patterns of interaction among entities) through time is thus presented as merely the differential outcome between particular moments in time, which essentially "freezes" change within a static and comparative conceptual framework that is more descriptive than it is explanatory (see Kerr 2002).

After the sudden and dramatic end of the Cold War, however, some theorists, recognizing the inability of traditional conceptual frameworks of security studies in international relations to explain change, have turned to an evolutionary approach for insight (Modelski 1996; Thompson 2001). An evolutionary framework places the very concept of change as central, does not treat agents and their identities as fixed and given, and does not "privilege a type of actor or a type of problem as the core foci" (Thompson 2001, p. 2). An evolutionary approach in international relations theory is not expected to solve all problems or lead to a "perfect" system of thought, but, rather, to provide new ways of looking at old problems.

While an evolutionary framework that focuses on change as an ongoing dynamic process holds advantages over static approaches to agency, the field of evolutionary international relations is still in its infancy about what an evolutionary paradigm should look like. One obstacle facing the research program is "What kinds of units . . . generate variations, are subject to selection pressures, adapt or fail to adapt, and demonstrate change that can be made more intelligible by an evolutionary perspective" (Rapkin 2001, 52)? The units of analysis problem in international relations theory is not new (Waltz 1959; Singer 1969), but in the context of an evolutionary framework a new wrinkle emerges in this issue because there is a set of criteria that any material configuration must meet to merit designation as an evolutionary "individual" that can participate in a process of selection (Gould 2002)—that is, when properties of a relevant individual interact with the environment in a causal way to influence the relative representation of future generations.

According to George Modelski, "the starting point for evolutionary analysis [in international relations] is the global political system viewed as a set of policies (or strategies). These policies may (conceptually) be carried by a variety of actors or agents. . . . [T]he emphasis at this point is not on actors . . . but on the policies themselves viewed as sets of instructions, or programs. The instructions embodied in global politics provide the basis for the standard operating rules, or routines, of the global political system" (Modelski 1996, 331).

This quote implies that the objects of selection in evolutionary international relations studies are fundamentally units of information. At bottom, then, evolution boils down to a competitive game to maximize representation of one variant version of information over another (Eldredge 1995). Given the intellectual roots of international relations theory and its emphasis on actors (human beings, states, groups of states, etc.), this view may seem vastly oversimplified (or perhaps even logically incoherent), because it effectively reduces actors or agents to mere vehicles that house information. When we consider the character of selection as a causal economic process, however, it becomes evident that this conceptualization is incomplete: it confuses a need for measuring the results of selection by counting differential increase of some informational attribute with the mechanism that produces relative "reproductive" success.

Biologists, too, are not immune to these conceptual errors. The unit of selection problem in biology has been an intellectual minefield. Confusion, it seems, has stemmed largely from the ways in which biological phenomena have been conceptualized (Hull 1988a). It is probably fair to assume then that the fledgling field of evolutionary international relations is bound to make some of the same conceptual mistakes. In this chapter, my comments are directed mainly toward efforts in the developing "macrolevel" evolutionary world politics research program in international relations (Modelski 1996, 2001). My comments about the selection process, however, are of a general nature, such that they also apply to other developing research programs in the evolutionary international relations paradigm (e.g., the biobehavioral point of view; see Falger 2001; Thayer 2004).

This chapter, therefore, drawing mainly on lessons learned by biologists in their own intellectual struggles, attempts to strengthen the conceptualization of selection in evolutionary international relations. I also briefly address the issue of ontology in evolutionary international relations, that is, the question of whether large-scale social entities actually exist with the capacity to act as causal agents (evolutionary individuals) in a process of selection. It is not my intention, given the magnitude of the issue, to answer this question explicitly for any social entity, but rather to outline some of the conceptual tools needed for answering it. I hope that this contribution will act as a springboard for effective communication between disciplines currently developing along orthogonal paths.

## What Is Selection?

A helpful starting point is to ask how to identify a level of generality that is not too broad so that everything becomes a selection process or too narrow so that each type of selection becomes unique (Hull et al. 2001). For Darwin, selection, due ultimately to competition among entities, proceeded in two steps: the production of variation (regardless of how it arises) and the sorting or nonrandom elimination of variation by selection according to the differential performance of entities in which this variation is expressed (Mayr 2004; Vermeij 2004). Darwin's theory also depends on the existence of heritable differences between individuals if change is to be of an evolutionary nature. This is one reason why Darwin was able to develop so many insights about natural selection even though he did not have the slightest idea about how the mechanism of inheritance worked. In short, entities selected are those that have traits that best fit the challenges and opportunities of their environment (Vermeij 2004). In other words, in the favored metaphor of Darwin's day, selection is basically a sieve—what worked better than what is differentially represented in the next generation.

Richard Dawkins (1983), building on the important work of especially Campbell (1965) and Lewontin (1970), argued that Darwin's principles of variation, selection, and inheritance (regardless of how information is transmitted, including Lamarckian inheritance of acquired characters through social learning and imitation) are fundamental to life everywhere. It is worth stressing though, that the logic of the principle of selection is so straightforward that it can hardly be questioned at all (Mayr 2004). This is one reason why Darwin's theory has been applied to fields of study as varied as cosmology (Chaisson 2001), economics (Hodgson 2002; Hodgson and Knudsen 2004), and the sociology of science (Hull 1988b). In fact, natural selection, as Darwin conceptualized it, is actually a special—some would say emergent (Depew and Weber 1996)—selection process in which the components are replicating. There is thus more than biological analogy involved here (Hodgson 2002). The set of "universal Darwinian principles"—variation, selection, and inheritance (Dawkins 1983)—is abstract enough to serve as a conceptual framework for the analysis of change in all systems composed of living entities, including social systems.

It is important to stress at this point that this approach does not invoke biological reductionism (Hodgson 2002), in which everything can be explained in biological terms. Instead, as Geoffrey Hodgson (2002, 270) has pointed out, "there is a core set of general Darwinian principles that, along with auxiliary explanations specific to each scientific domain, may apply to a wider range of phenomena."

## Selection and Environment

In pointing out the obstacles to applying selection theory to an evolutionary international relations program, David Rapkin (2001, 54) suggested "logical problems arise immediately" when applying Darwin's theory to the evolution of global politics, because objects of selection do not just respond to their environment, they also influence the environment. That is, part of the security environment is not exogenous to the social units (e.g., states) that exist within it because the environment may contain, among other things, "more powerful units [that] are able to shape the environment so as to alter the constraints and opportunities it posed for themselves and for other units. For example, hegemonic states exercise structural (or meta-) power to set the agenda of global issues, impose ordering principles, promote certain norms and types of collective action, and more generally, to determine the 'rules of the game' for other states and for other types of social units."

This quote from Rapkin's article, however, could just as easily have been written about organisms in nature. Organisms in nature not only determine the aspects of the outside world that are relevant to them, but they actively create or construct the world around them (Lewontin 2000). Geerat Vermeij (2004, 3), in his discussion of the fundamental principles that govern the evolutionary process, wrote: "Dominants exercise disproportionate influence and accumulate disproportionate power and wealth. Through top-down control, they affect not only the characteristics and distribution of other members of the economy, but they define the structure and workings of the economy as a whole." Thus the claim that the environment of an organism is causally independent of the organism is mistaken. The environment is not simply autonomous to the organism as Darwin's theory imposed (and Rapkin was correct in pointing out), but a reflection of the biology of an organism. In short, entities, at all scales of economic life, from the cell to ecosystems, from the firm to states, influence their own evolution by being both the object of selection and the creator of the conditions of that selection.

## Units of Selection

Every field must struggle with understanding its most basic units and levels. The issue of what entities act as agents or units of selection depends crucially on the nature and logic of the theory of selection itself. The issue is not merely semantic or philosophical. As mentioned in the introduction, the central issue of causal agency in selection has generated much confusion in biology (Hull 1988a; Ghiselin 1997; Gould 2002), arising not so much from disagreement about empirical matters, but from conceptual problems about the meaning of causality in selection. It is not surprising then to find the seeds of a similar problem developing in evolutionary international relations studies. Although it is generally acknowledged (Rapkin 2001; Thompson 2001) that one advantage of an evolutionary approach is its flexibility in the units of analysis, the ontological reality of these entities as causal agents in a process of selection largely has been taken for granted. Individual human beings, cities, states, nations, nongovernmental organizations, firms, strategies, policies, behaviors, genes, ideas, norms, institutions, and "societies of states," among others have all been suggested by various authors as candidate units of selection that may be "examined fruitfully from an evolutionary standpoint" (Rapkin 2001, 53).

Similarly, there also are all sorts of units in biological systems—lengths of RNA, genes, chromosomes, antibodies, cells, organisms, groups of organisms (e.g., colonies), species, clades, communities, ecosystems—that have been treated as units of selection at one time or another (sometimes inappropriately). This list includes both genealogical (replicative) and ecological (interactive or economic) entities. As biologists have come to appreciate, however, genealogical and ecological entities play very different roles in the process of evolution by selection (Eldredge 1985), and failure to disambiguate the two often leads to ontological confusion about causal agency in selection. An understanding of what historical entities are and are not units of selection— a task that is not easy, to say the least—is thus an important starting point. It is in this sense that I briefly outline the units of selection problem in biology as a guide to security theorists in dissecting the individuation problem.

Selection, in any domain, first and foremost, must operate among "evolutionary individuals" (and, by extension, potentially at several levels in a hierarchy (or hierarchies) of units, each properly construed as an individual (Gould 2002). Darwin's key insight was that selection can operate only by the differential reproductive success of individuals. It is thus critical that we specify a set of criteria that any material configuration must meet to merit designation as an individual. I use the term "individual" to refer to a level or unit of selection (Hull 1980). According to Ghiselin (1974, 536), "In [metaphysics or] logic, 'individual' is not a synonym for 'organism.' Rather, it means a particular thing"—even if some entities, such as species, do not seem like individuals from the human perspective. I also will closely follow Stephen Jay Gould's (2002) helpful discussion of what an individual is in evolutionary terms, although I place more weight, as do Wilson and Sober (1994), on the criterion of functional interdependence.

According to Gould (2002), three sets of criteria, ranging from conventions of ordinary language to specific requirements of Darwinian logic, define individuality. First, to be an individual, in the vernacular sense of the word, a material entity must have (1) a discrete and definable beginning, or birth; (2) an equally discrete and definable ending, or death; and (3) sufficient stability to merit continuous recognition as the same "thing."

Second, these vernacular individuals must then manifest the essential property that permits them to function as an evolutionary individual with the capacity to act as a causal agent in a process of Darwinian selection. Mere vernacular individuality does not suffice for identification as a causal actor in universal Darwinian theory. A principle of inheritance must prevail. Successive variations must in some sense be retained and then passed on (Hull et al. 2001). Without such continuity in heredity, differential reproductive success will impart no Darwinian advantage. In other words, evolutionary individuals must be able to pass—differentially and in a heritable manner—their favorable properties into future generations. Although selection may produce change, evolution does not occur without heritability. This does not imply, however, that units of selection must pass "faithful" copies of themselves in the next generation. Units of selection only need to increase the relative representation of their heritable attributes in the next generation (Sober and Wilson 1994; Gould 2002).

Third, evolutionary individuals must function as interactors with the environment (Brandon 1990). Interaction with environment defines economic activity, causing replication to be differential (Hull 1980). We must, therefore, be able to formulate a testable causal scenario about why certain heritable characteristics yield increased reproductive success. Wilson and Sober (1994) further suggested that functional interdependence (a shared fate) among parts (or components) that interact together is a main criterion for identifying units of selection. Gould (2002) objected to this criterion, not in the "invocation of functionality itself" as a property involved in our ordinary concept of sufficient stability, but with Wilson and Sober's stress on "organismlike" properties that emphasize direct modes of interaction (e.g., active competition with the environment). Gould preferred a looser or broader concept of interaction that also included modes of indirect interaction (see also Williams [1992, 25] for a similar view). This looser concept opens the door for Gould and others to view individuals, such as species (the "poster-child" of the individuality thesis in biology), as adequate interactors in a process of selection. (I side with those that do not grant species an economic role as units of selection for reasons that are beyond the scope of the present chapter; see Eldredge [1995], for a more in depth treatment of this contentious issue.) The important point here, however, is not what mode of interaction is acceptable, but that it is agreed that units of selection must be interactors.

What counts as an interactor is thus essential to the advancement of an evolutionary research program in international relations; this point can hardly be overstated. In biology, for instance, Richard Dawkins's (1982) view of a unit of selection, which seems to be the primary inspiration for supporters of an evolutionary approach in international relations (Florini 1996; Modelski 1996), is likely to mislead one into treating interactors ("vehicles" in Dawkins' terminology) as merely passive entities (Ghiselin 1987). This philosophical perspective results in a deemphasis on the identity of the causal interactors in the selection process. Everything flows from competition among replicators—anything in the universe of which copies are made, such as genes and routines—to leave more copies of information to the next generation. Adherence to this view forces a reconceptualization of the domain of evolutionary causality; instead of focusing on the identity of interactors, the motivating question becomes what are the replicators in world social organization? This reductivist (what Niles Eldredge calls the ultra-Darwinian) influence from biology is also evident in work by Florini (1996), in which change in international relations "is the result of competition among norms that are reproduced at different rates and that thus come to have different frequencies in the population of states (p. 369)." Florini (1996) suggested that we cannot look at states or firms (potential interactors), because selection rarely works directly on these economic entities. "By following the evolutionary journey from the perspective of the gene, or norm, rather than from the perspective of the organism, we get much better insight into factors accounting for norm change (p. 371)." But replicators simply are not units of selection or, for that matter, causal agents, unless they also happen to be interactors (Gould 2002). For instance, in biology, genes (replicators), or any other genetic elements that can propagate themselves within the genome independent of effects at other levels, can also function as interactors and therefore as agents of selection (see Wilson and Sober [1994, 592] and Gould [2002, 689] for details of specific examples). It is interactors that survive or fail to survive, reproduce or fail to reproduce, not units of information. Units of information are not "visible" to selection and therefore can not serve as a target (Mayr 2004). What is visible to selection is the phenotype of the interactor that screens off the underlying units of information (Brandon 1990). Replicators are important in evolution, but in a different role as items for bookkeeping (Gould 2002). Leaving out reference to interaction leaves out reference to the entities "keeping the books" (Hull 1988a).

If we think of evolution as an economic process in which entities and their replicators are embedded, our emphasis shifts to the complex interplay of competitive and cooperative interactions and, more importantly, to their economic consequences (Corning 2005). In addition, if causality resides in interactors, and interactors at several levels can be established as legitimate evolutionary individuals (simply due to a consequence of having had a history), then a hierarchical theory of selection becomes unavoidable as a coherent logical structure.

## Multilevel Selection

Multilevel (or hierarchical) selection theory takes as its starting point the idea that selection can operate simultaneously at different levels of a hierarchy, for instance, the ecological hierarchy in biology (organism, population, community, ecosystem). Hierarchical selection can be defined as differential proliferation of relevant evolutionary individuals based on causal interaction of their properties with surrounding environments (Gould 2002). The existence of evolution on multiple levels, however, does not necessarily involve replicators at different levels (Brandon 1990). Deciding what the replicators are is not central to the units of selection problem, though establishing that traits at a given hierarchical level are heritable is essential for evolutionary change to occur (Sober and Wilson 1994; Godfrey-Smith 2000; Gould 2002). Thus, evolution in world politics, theoretically, can be affected by selection at more than one level. This implies then that it is a conceptual error to ask in international relations what the level of selection is in a given context or to expect a "clear best choice" to emerge. In general there need be no single answer.

An important concept to mention in multilevel selection theory is sorting. Vrba and Gould (1986, 217) defined sorting as "a simple description of differential representation; it contains, in itself, no statement about causes." At its core, selection causes sorting, but adequate evidence for selection can not be just the simple observation of sorting, because a multilevel theory of selection adds other potential causes of sorting: Levels can interact positively, negatively, or orthogonally (independently), and events at one level may propagate as effects to others. Causation may flow from lower levels to higher, or in the reverse (upward and downward causation [Vrba 1989; Gould 2002]). For instance, in biology, sorting at the species level is often causally reducible to the lower organismal level (Vrba 1989). In this example, no distinct causality emerges at the species level, because the characters responsible for selection belong to organisms and thus are merely transferred as an effect to the species level by upward causation from a lower level. If more than one hierarchical level simultaneously affects another, we may also not be able to disentangle the differences, leaving us with an account of consequences, rather than causes (Gould 2002).

I have not attempted to answer the empirical question of what the efficacy of higher-level selection is in evolutionary world politics. Such a line of inquiry is beyond my expertise and simply not possible at this time; the development of a nested (inclusive) hierarchy of entities, not to mention the ontology of higher-level evolutionary individuals that can participate as causal agents in a selection process, remains contentious. Are large-scale social organizations, such as states (the axiomatic unit of international relations), mere collections of entities that share some properties; that is, are they idealizations—arbitrarily defined heuristic constructs—rather than real entities themselves? This view seems commonly held in the social sciences, from my reading of the literature (see Campbell 1994; Khalil 1995). Or are social systems ontological entities in their own right, even though the self-organization of such entities emerges largely from entities interacting at a lower level (e.g., self-interested human actors)? In the former case, selection is unimportant at higher levels of organization (e.g., state level), because these emergent levels can be explained in terms of selection at lower levels. Larger-scale entities thus play no causal role themselves in the evolutionary process and are relegated to mere epiphenomena in which the competitive drama unfolds. Attempts to incorporate hierarchical selection theory in the evolutionary world politics research program (e.g., Devezas and Modelski 2003) have not rigorously addressed this crucial issue and thus do not necessarily qualify as valid statements of higher-level selection.

Rather, I have tried to shed light on why this set of issues must be addressed in a more systematic fashion. Perhaps this line of reasoning is what Thompson (2001, 6) had in mind: "The question is whether it is feasible analytically to ignore other evolutionary developments while choosing to focus only on one level at a time." It should be clear from the above discussion that if multiple levels of selection exist (which is still an open question in evolutionary international relations), study at one level of analysis can not proceed logically with little concern for what is happening at other levels; that is, it would never be "feasible" to ignore these other levels because the direction of selection may be different at different hierarchical levels.

An important point of this section, as with previous sections, however, is that when there is selection at higher levels, these entities must be economic interactors because they alone are causal agents in selection (Brandon 1990; Gould 2002). I must emphasize though that the mere conceptual coherence of higher-order selection does not necessarily mean that such potentialities have been realized. It is never appropriate to expand the levels of selection without good reason (Ghiselin 1997).

## Transitions in Levels of Selection

Over the course of the past 13,000 years the trend in human social evolution has been one of the replacement of smaller, less complex units by larger, more complex ones (Diamond 1997). This hierarchy of organization is itself the product of evolution—entities further up the hierarchy, such as cities and states, have obviously not been there since the beginning. These discrete steps, characterized by a higher level of complexity, are termed "transitions." Ideally, then, we would like an evolutionary theory that also explains the emergence of higher-level entities. Cederman (1997, 4) stated that "it is precisely the issue of unstable agency that poses the most serious conceptual problems to existing IR (international relations) approaches. Almost without exception, these perspectives assume the actors are exogenously given." . . . "Accounting for their origin, development, and disappearance is a necessary ingredient of any long-term explanation of world politics" (p. 16). The challenge then is to understand these transitions in Darwinian terms. In short, we want to know how the hierarchy—the complexification of social organization—evolved in the first place. The logic behind this complexification appears obvious: a higher-order system can be erected only from simpler systems after these building blocks have evolved themselves. The levels of selection question thus becomes not only the problem of what hierarchical level(s) selection now acts, but the problem of determining how the levels in the hierarchy evolved (Okasha 2006).

Peter Corning (2005) has suggested that synergistic (cooperative) effects have played an important role in the emergence of more complex systems, both in nature and in the human realm (see also Vermeij 2004). Synergy engenders more complex forms with a competitive edge. Corning hypothesizes that it was the selective advantages of cooperative interactions of various kinds that have facilitated the evolution of complex, functionally organized biological and social systems. Within a given environmental context, it is the benefits of synergistic innovations that lead to long-term evolutionary changes in the direction of greater complexity (Corning 1996, 2005). For instance, under Corning's view, if an effective global security system emerges, it will be the product of a cooperative effort in which participants find it advantageous to place control in an entity that is shared in common. It is clear, however, that the world presently is no where near such a collective synergy, or even whether it can be supported, given that the most powerful institutions will voluntarily have to relinquish their power to make it work. In short, cooperative interactions that produce positive functional consequences to the benefit of the new larger entity, however they may arise, can lead to higher-level units of selection (see also Buss [1987] for discussion of synergy and the evolution of hierarchical levels in biology, such as the origin of the highly integrated economy of the eukaryotic cell).

Self-organization theories (Geiger 1988; Kauffman 1993), which suggest that much of the order found in complex systems may spontaneously emerge (autocatalytic), may seem to oppose Corning's selectionist theory. Selection in Kauffman's paradigm plays a limited (or at least supporting) role (see also Depew and Weber 1996). While self-organizing processes may facilitate the process of constructing complex systems, the results are always subject to the final editing of selection (Corning 2005; Wilson 2005). Self-organization thus persists because its functional effects work sufficiently well.

As a final point in this section, it is worth mentioning that the very concepts that I outlined previously to understand multiple levels of selection become problematic when we ask how the levels in a hierarchy evolved. For instance, Okasha (2006) has pointed out that the replicator and interactor conceptualization of selection is not well suited for studying transitions in organizational complexity. The cohesiveness of an interactor is itself a highly evolved property. To understand how cohesiveness (functional interdependence or shared fate) evolved, it obviously becomes problematic to build this notion into the very concepts used to describe evolution by selection (Okasha 2006).

## Selection, Predictability, and Directionality

According to Gaddis (1992, 53), the major international relations approaches have all had, with "strikingly unimpressive results," the principal objective of predicting or forecasting the future course of world events. The acid test of the success of an analytical construct in international relations is supposedly whether it can a priori predict future change. This leads to the logical question of whether "it should be required that any evolutionary theory of global politics specify the direction(s) in which . . . change is headed" (Rapkin 2001, 56).

Selection is a historical, cumulative process, with context-dependent factors playing a crucial role at every step. In other words, it is an a posteriori phenomenon (Mayr 2004). It is not a deterministic (lawlike) process in which some external or internal force ultimately controls the direction of the process, such that a priori predictions are possible. Thus, although Modelski (1996, 2001) has adopted a more explicitly evolutionary paradigm to structural changes in international relations, his envisioning of the global process as "unfolding" according to an "inner logic" is still problematic (quoted words cited in Corning 2005, 144). The historically contingent nature of selection cannot be overemphasized. The take-home message is that we can not use past trends to "safely" predict the future (Corning 2005). Nobody can know where the trend in complexification of social organization (Diamond 1997), for instance, is specifically headed—evolution is not a movement toward any specific predestined goal.

The contingent nature of all dynamic change has led some observers in biology to deny that history can reveal any consistent long-term patterns and trends in nature (Gould 1996, 2002). But, others, notably Geerat

Vermeij (1999)—without any teleological notion of implied progress, value, or design—has stressed that although "unpredictable events still intercede, and starting conditions still affect when, where, and how subsequent events unfold, with the result that the precise time course of history and the exact identity and nature of the participants remain unpredictable, some universal rules governing participants and their actions enable us to discern underlying temporal patterns [directionality] that are predictable and scientifically testable" (p. 243). Thus, although selection is a probabilistic process, foregoing any unequivocal predictions of the future, it would be foolish to use this reason as an excuse to turn our backs on history for security insights. Evolutionary studies highlight what has worked and what has not under every imaginable circumstance, which adaptive hypotheses—continually tested by selection for predictive value—have preserved flexibility and opportunity and which have led to rigidity and constraint (Vermeij 2004; Chapter 3).

## Concluding Statements and Further Research

Biologists agree on many things, among them the ubiquity of change in nature. But important conceptual issues in selection theory still remain contentious. I cannot say it any better than Michael Ghiselin did in presenting his views on the units of selection controversy in biology in the preface to his book *Metaphysics and the Origin of Species* (1997, x): "It is downright impossible to discuss difficult and contentious issues without disagreeing with all sorts of people." The application of one set of evolutionary models in international relations is apt to yield different results than another alternative. These admissions do not, however, in any way, undermine the generality of the universal Darwinian principles that I have outlined in this chapter: variation, selection, and inheritance. These principles are abstract enough to serve as a framework for the analysis of evolutionary change in all organized systems. In fact, selection seems so universal that it might be better to ask how it comes about and what its consequences are, rather than merely establishing that it took place (Vermeij 2004).

With these principles as an anchor, I have attempted to draw attention to the lessons learned by biologists in their exchanges over the units of selection problem in order to strengthen the conceptualization of selection in the emerging evolutionary international relations research program. A particularly urgent task seems to be the further refinement of the ontological status of large-scale social organizations as evolutionary individuals, instead of taking the existence of these entities for granted. In addition, selection is a causal process, and units or agents of selection must be defined as overt actors (interactors) in the mechanism, not merely as preferred items in the "calculus of tabulating results" (Gould 2002). Envisioning replicators as the fundamental units of selection confuses what gets selected with what gets affected by selection.

If interactors at several levels can be established as legitimate evolutionary individuals, then a hierarchical theory of selection becomes unassailable. If a theory of hierarchical selection develops in evolutionary international relations, keeping in mind that there must be good reason to expand the levels of analysis, it must be stressed that levels of causality are bonded in interaction, with selection at one level enhancing, counteracting, or just independent to selection at an adjacent level. The levels may also operate differently despite unifying principles, such as selection acting at each level. Vague statements about the efficacy of higher levels of selection must be avoided, because they can only impede our understanding of the theory of selection in international relations.

On one hand, I agree entirely that an evolutionary approach to international relations holds obvious advantages over traditional theoretical frameworks that effectively "freeze" change. For Heraclitus of Greek antiquity, as for Darwin, change is inevitable. A deeper understanding of change seems even more pressing today because, whether we like it or not, connections between nations and peoples are growing in all aspects of our lives thanks to the globalization of markets for goods, services, and capital. These developments are inevitably creating new challenges that alter the adaptive landscape of the global security environment, such as today's threat of global terrorism. The capacity of entities to adapt to the changing structure of today's security environment remains key to survival in the human realm of international relations, as surely it does in nature. But, like organisms in nature, entities in international relations are more than just hypotheses of their environment; they also create their surroundings. Moving beyond conceptual frameworks of security that depend on the perilous balance of power between unproblematic state actors seems a necessary step forward.

On the other hand, it is also easy to be a critic. When you think about any complex issue, it quickly becomes apparent why this or that approach to its study might be misleading or inappropriate. It is more difficult to suggest how we can, in practice, do better. My attempts here, while pointing out the limitations of one approach, have only added to the alternative visions in the universe of possibilities of what an evolutionary paradigm of structural change in international relations might look like. Ultimately, the advantage of an evolutionary approach in international security studies depends on whether it leads to better theories than currently are available: we are still a long way from evolutionary international relations. I am optimistic, however, that historical approaches to international relations that are grounded in evolutionary theory will not only help teach us about the world we live in now and how we got here, but also what might shape our future.

ACKNOWLEDGMENTS

I am grateful to Rafe Sagarin and Terry Taylor for the invitation to contribute to the NCEAS working group and this volume. I also thank Gary Vermeij, Dominic Johnson, and Warren Allmon for their helpful reviews and/or discussions in the early formative stages of this chapter. A Donnelley Environmental Fellowship at Yale University partially supported this work.

REFERENCES

Brandon, R. N. 1990. *Adaptation and environment*. Princeton, NJ: Princeton University Press.

Buss, L. 1987. *The evolution of individuality*. Princeton, NJ: Princeton University Press.

Campbell, D. T. 1965. Variation and selective retention in socio-cultural evolution. In *Social change in developing areas: A reinterpretation of evolutionary theory*, ed. H. Barringer, G. I. Blanksten, and R. W. Mack, 19–48. Cambridge, MA: Schenkman.

Campbell, D. T. 1994. How individual and face-to-face-group selection undermine firm selection in organizational evolution. In *Evolutionary dynamics of organizations*, ed. J. A. C. Baum and J. V. Singh, 23–38. New York: Oxford University Press.

Cederman, L.-E. 1997. *Emergent actors in world politics: How states and nations develop and dissolve*. Princeton, NJ: Princeton University Press.

Chaisson, E. J. 2001. *Cosmic evolution: The rise of complexity in nature*. Cambridge, MA: Harvard University Press.

Corning, P. A. 1996. Synergy, cybernetics and the evolution of politics. *International Political Science Review* 17: 91–119.

Corning, P. A. 2005. *Holistic Darwinism: Synergy, cybernetics, and the bioeconomics of evolution*. Chicago: University of Chicago Press.

Dawkins, R. 1982. *The extended phenotype: The long reach of the gene*. New York: Oxford University Press.

Dawkins, R. 1983. Universal Darwinism. In *Evolution from molecules to men*, ed. D. S. Bendall, 403–425. New York: Cambridge University Press.

Depew, D. J. and B. H. Weber. 1996. *Darwinism evolving: Systems dynamics and the genealogy of natural selection*. Cambridge: The MIT Press, p. 588.

Devezas, T., and G. Modelski. 2003. Power law behavior and world system evolution: A millennial learning process. *Technological Forecasting and Social Change* 70: 819–859.

Diamond, J. 1997. *Guns, germs, and steel: The fates of human societies*. New York: W.W. Norton and Company.

Eldredge, N. 1985. *Unfinished synthesis: Biological hierarchies and modern evolutionary thought*. New York: Oxford University Press.

Eldredge, N. 1995. *Reinventing Darwin: The great debate at the high table of evolutionary theory*. New York: John Wiley and Sons.

Falger, V. S. E. 2001. Evolutionary world politics enriched: The biological foundations of international relations. In *Evolutionary interpretations of world politics*, ed. W. R. Thompson, 30–51. New York: Routledge.

Florini, A. 1996. The evolution of international norms. *International Studies Quarterly* 40: 363–389.

Gaddis, J. L. 1992. International relations theory and the end of the Cold War. *International Security* 17: 5–58.

Geiger, G. 1988. On the evolutionary origins and function of political power. *Journal of Social and Biological Structures* 11: 235–250.

Ghiselin, M. T. 1974. A radical solution to the species problem. *Systematic Zoology* 23: 536–544.

Ghiselin, M. T. 1987. Bioeconomics and the metaphysics of selection. *Journal of Social and Biological Structures* 10: 361–369.

Ghiselin, M. T. 1997. *Metaphysics and the origin of species.* Albany: State University of New York Press.

Godfrey-Smith, P. 2000. The replicator in retrospect. *Biology and Philosophy* 15: 403–423.

Gould, S. J. 1996. *Full house: The spread of excellence from Plato to Darwin.* New York: Harmony Books.

Gould, S. J. 2002. *The structure of evolutionary theory.* Cambridge, MA: Belknap Press of Harvard University Press.

Hodgson, G. M. 2002. Darwinism in economics: From analogy to ontology. *Journal of Evolutionary Economics* 12: 259–281.

Hodgson, G. M., and T. Knudsen. 2004. The firm as an interactor: Firms as vehicles for habits and routines. *Journal of Evolutionary Economics* 14: 281–307.

Hull, D. L. 1980. Individuality and selection. *Annual Review of Ecology and Systematics* 11: 311–332.

Hull, D. L. 1988a. Interactors versus vehicles. In *The role of behavior in evolution,* ed. H. C. Plotkin, 19–50. Cambridge, MA: MIT Press.

Hull, D. L. 1988b. *Science as a process.* Chicago: University of Chicago Press.

Hull, D. L., R. E. Langman, and S. S. Glen. 2001. A general account of selection: Biology, immunology, and behavior. *Behavioral and Brain Sciences* 24: 511–573.

Kauffman, S. A. 1993. *The origins of order: Self-organization and selection in evolution.* New York: Oxford University Press.

Kerr, P. 2002. Saved from extinction: Evolutionary theorizing, politics and the state. *British Journal of Politics and International Relations* 4: 330–338.

Khalil, E. L. 1995. Organizations, naturalism, and complexity. *Review of Social Economy* 53: 393–419.

Lewontin, R. C. 1970. The units of selection. *Annual Review of Ecology and Systematics* 1: 1–18.

Lewontin, R. 2000. *The triple helix: Gene, organism, and environment.* Cambridge, MA: Harvard University Press.

Mayr, E. 2004. *What makes biology unique?* New York: Cambridge University Press.

Modelski, G. 1996. Evolutionary paradigm for global politics. *International Studies Quarterly* 40: 321–342.

Modelski, G. 2001. Evolutionary world politics: Problems of scope and method. In *Evolutionary interpretations of world politics,* ed. W. R. Thompson, 16–29. New York: Routledge.

Okasha, S. 2006. The levels of selection debate: Philosophical issues. *Philosophy Compass* 1: 1–12.

Rapkin, D. P. 2001. Obstacles to an evolutionary global politics research program. In *Evolutionary interpretations of world politics,* ed. W. R. Thompson, 52–60. New York: Routledge.

Singer, J. D. 1969. The level-of-analysis problem in international relations. In *International politics and foreign policy: A reader in research and theory,* ed. J. N. Rosenau, 20–29. New York: The Free Press.

Sober, E., and D. S. Wilson. 1994. A critical review of philosophical work on the units of selection problem. *Philosophy of Science* 61: 534–555.

Thayer, B. A. 2004. *Darwin and international relations: On the evolutionary origins of war and ethnic conflict.* Lexington: University Press of Kentucky.

Thompson, W. R. 2001. Evolving toward an evolutionary perspective. In *Evolutionary interpretations of world politics,* ed. W. R. Thompson, 1–14. New York: Routledge.

Vermeij, G. J. 1999. Inequality and the directionality of history. *American Naturalist* 153: 243–253.

Vermeij, G. J. 2004. *Nature: An economic history.* Princeton, NJ: Princeton University Press.

Vrba, E. S. 1989. Levels of selection and sorting with special reference to the species level. *Oxford Surveys in Evolutionary Biology* 6: 111–168.

Vrba, E. S., and S. J. Gould. 1986. The hierarchical expansion of sorting and selection: sorting and selection cannot be equated. *Paleobiology* 12: 217–228.

Waltz, K. N. 1959. *Man, the state and war: A theoretical analysis.* New York: Columbia University Press.

Whitehead, A. N. 1925. *Science and the modern world.* New York: Macmillan.

Williams, G. C. 1992. *Natural selection: Domains, levels, and challenges.* New York: Oxford University Press.

Wilson, D. S. 2005. Natural selection and complex systems: A complex interaction. In *Self-organization and evolution of social systems,* ed. C. K. Hemelrijk, 151–165. Cambridge: Cambridge University Press.

Wilson, D. S., and E. Sober. 1994. Reintroducing group selection to the human behavioral sciences. *Behavioral Brain Sciences* 17: 585–654.

Part Four

# EVOLUTION'S IMPRINT

## PSYCHOLOGY AND THE ROOTS OF TERRORISM

# Chapter 7

# MILITANTS AND MARTYRS

Evolutionary Perspectives on Religion
and Terrorism

RICHARD SOSIS AND CANDACE S. ALCORTA

The main argument of this chapter is that evolutionary studies of religion are vital for understanding the proliferation, patterns, and logic of current trends in terrorist activity. The importance of this message is becoming increasingly evident. As Simon and Benjamin (2000, 59) prophetically warned before 9/11, the threat of terrorism will "intensify, because the old paradigm of predominantly state-sponsored terrorism has been joined by a new, religiously motivated terrorism." Indeed, in recent years there has been a rise in the proportion of terrorists motivated by religious concerns, and there is a significant correlation between religious motivation and lethality (Hoffman 1998). Simon and Benjamin argue that state-sponsored terrorism is somewhat constrained since states do not want to undercut their claims of legitimacy and alienate potential supporters who would revile indiscriminate violence against civilians. In contrast, the new religiously motivated terrorists "want a lot of people watching and a lot of people dead" (Simon and Benjamin 2000, 71). Why is religiously motivated terrorism becoming more common? And why is it more deadly? What governmental policies can stem religiously motivated violence? We believe evolutionary research on religion can provide novel insights and timely answers to these pressing questions.

In this chapter we first examine the relationship between religion and terrorist activity and clarify religion's role in causing, motivating, and facilitating terror. Next we present recent work on the evolution of religion and show how this research can help us understand current patterns of terrorism. The chapter concludes with future research questions and policy implications derived from our evolutionary approach.

## What Is the Relationship between Religion and Terrorism?

While the literature on terrorism offers widely divergent opinions on almost every topic, one matter is clear: not all terrorists are religious fanatics, or even religious. Nonetheless, it is also apparent that religiously motivated terrorism is on the rise and it is alarmingly more lethal than other forms of terrorism. One possible explanation for the lethality of religious terrorism is that it is more closely linked to suicide terrorism, which accounts for only 3% of all terrorist activity from 1980 through 2003, but 48% of the fatalities (Pape 2005, 6). Some researchers have indeed argued that religious motivations are more commonly associated with suicide terrorism (Hoffman 2003; Berman and Laitin 2005), although others have rejected this connection (Bloom 2005; Pape 2005). In support of the latter position, it has been widely observed that the Liberation Tigers of Tamil Eelam (LTTE), a Marxist-Leninist group demanding independence from Sri Lanka, were the world leaders in suicide terrorism prior to the current insurgency in Iraq. Furthermore, Pape's (2005, 210) analysis of data on suicide terrorism from 1980 to 2003 found that less than half of the suicide bombings documented during this period were religiously motivated. However, even if religiously motivated suicide bombers are not more common than their secular counterparts, they may be more successful in causing fatalities, as Berman and Laitin's (2005) data suggest. Below we argue that religion can institutionally and individually facilitate suicide attacks and other forms of terrorism, and that specific evolved elements of religion differentially contribute to these effects. We also contend that these evolved elements of religion are employed by both secular and religious terrorists alike in order to achieve their goals and create a committed and cohesive following.

The media may be responsible for the popular belief that religion and terror are strongly associated. Terrorists, especially suicide terrorists, are often depicted in the media as delusional religious fundamentalists, hopelessly brainwashed and out of touch with reality. The picture that is emerging from recent research, however, is far different. For example, Berrebi (2003) has shown that Palestinian suicide bombers have above average education and are economically better off than the general population. Krueger and Maleckova (2002) also demonstrate that poverty is not a predictor of participation in political violence or support for terrorism. Moreover, Sageman (2004) found no evidence of psychopathology in an international sample of Muslim terrorists. Leaders of terrorist organizations are clear that recruits may not be depressed or suicidal. As one spokesman for the Palestinian Islamic Jihad explains, "to be a martyr bomber you have to want to live" (Richardson 2006, 117). Terrorists themselves point out that even suicide bombers have plenty of hope, otherwise there would be no point in killing themselves (Atran 2003).

If terrorists in general, and suicide bombers in particular, are not crazed religious zealots, what then is the relationship between religion and terrorism? Various researchers have argued that terrorists have political, not religious goals (Juergensmeyer 2003; Bloom 2005; Pape 2005). Religion is not the root cause of conflicts but is rather a tool used by terrorists to achieve their goals. Recast in evolutionary terms, religious beliefs, rituals, and institutions are proximate mechanisms that facilitate otherwise improbable behavioral outcomes. Here we review four main reasons why religion serves as an effective mechanism for terrorists.

## Framing the Conflict

Juergensmeyer (2003) argues that while religion is not the cause of most conflicts involving terror, religion is the means by which terrorists translate a local political struggle into a cosmic war. In other words, terrorists often frame their disputes in religious rather than political terms. This has various advantages, most significantly in motivating others to sacrifice themselves for the cause. This transformation from political to religious struggle encourages actors to perceive that they are participating in something of divine significance that transcends individual self-interest. Among Sikh militants in the Punjab, Juergensmeyer describes joining the struggle as "motivated by the heady sense of spiritual fulfillment and the passion of holy war" (2004a, 2). It is remarkable how successful contemporary terrorists have been in shaping world views so that they are consistent with their own views. Bin Laden, for instance, has been particularly successful in transforming his local grievance (getting U.S. troops off "Muslim" soil) into a cosmic clash between civilizations. The use of religion to transform local power struggles into cosmic conflicts benefits terrorist groups who may otherwise be viewed as economically and politically self-serving. In an age of instantaneous electronic communications, such religious framing of essentially local conflicts serves to broaden both the ideological and geographic base of terrorism. A second consequence of the religious framing of political conflicts is the extension of the horizon for victory. Terrorists perceive that they are fighting a cosmic war in divine time, thus eliminating incentives to "win" within one's own lifetime. Commenting on an interview with Hamas leader Abdul Aziz Rantisi, Juergensmeyer observes that "[i]n his calculation, the struggles of God can endure for eons" (2004b, 35).

## Moral Justification

Religion also facilitates terrorists' goals by providing moral legitimacy to their cause (Juergensmeyer 2004c). All contemporary world religions impose a moral framework upon their adherents, thereby enabling terrorists to present their conflicts in morally absolute dichotomies, such as good versus bad or righteous versus evil. While legitimizing ones' own cause, religions are particularly effective at demonizing those with opposing views.

The history of religion is replete with examples in which in-group passions are aroused and out-group hatreds are dangerously ignited. Indeed, one consistent predictor of suicide terrorism is a religious difference between the perpetrator and victim (Pape 2005). This occurs even when the terrorist group appears to have secular motivations, such as the LTTE, who are Hindus fighting a Buddhist majority. In Berman and Laitin's (2005) extensive sample of suicide terrorism, almost 90% of the attacks were aimed at victims of a different religion.

## Spiritual and Eternal Rewards

Religion not only provides a divine dimension and moral legitimacy to terrorist activity, it also defines the rewards that combatants can attain. After considering the benefits that Sikh militants attain, Juergensmeyer concluded that "[t]he reward for these young men was the religious experience in the struggle itself: the sense that they were participating in something greater than themselves" (2004a, 2). In addition to such spiritual rewards of transcendence, religion may also explicitly offer benefits in the afterlife that can rarely be matched in this world. The 9/11 hijackers all believed that they "would meet in the highest heaven" (Lincoln 2003, 98), which we can assume helped them rationalize their actions.

## Religious Symbols, Myths, and Rituals

Religion's most significant role in terrorism may be its incorporation of emotionally evocative and highly memorable symbols, myths, and rituals that serve to individually motivate and collectively unify diverse individuals under a common banner. All terrorist groups face the challenge of creating group commitment and individual devotion to a common cause. Anthropologists have long noted that fundamental "faith-based" elements of religion— symbols, myths, and rituals—foster this in-group commitment better than any other social institution. Not surprisingly, secular and religious terrorists alike maintain communal rituals and initiation rites that communicate an individual's level of commitment to the group (Atran 2003). For religious terrorists, cohesiveness is further fostered through powerful religious symbols, which "often become focal points in occupations involving a religious difference" (Pape 2005, 89). And of course, martyrdom itself means to sacrifice one's life for one's faith. Religion provides the rituals and symbols to both motivate and memorialize these local heroes, thereby affording them an otherwise unattainable status that is also eternal. Pape observes that "[s]uicide terrorist organizations commonly cultivate 'sacrificial myths' that include elaborate sets of symbols and rituals to mark an individual attacker's death as a contribution to the nation" (2005, 29).

## Terrorism and Evolutionary Research on Religion

Evolutionary studies of religion further our understanding of these elements of terrorism and provide important insights that can benefit the development of antiterrorist policies. There is unfortunately great misunderstanding among many social scientists concerning the merit of evolutionary analyses of religion. For example, Zuckerman claims that recent data on the high rates of atheism in many parts of the world "delivers a heavy blow" (2006, 13) to evolutionary theories of religion. Zuckerman mistakenly equates evolutionary theory with genetic determinism and falsely assumes that recent evolutionary research, which reveals psychological biases toward supernatural belief (e.g., Bering 2006), must be inaccurate since plenty of people appear to not exhibit these biases. Evolutionary theory does not endorse genetic determinism. On the contrary, evolutionary theory assumes that environmental input during ontogeny is critical for the expression and adaptive functioning of many traits, including religious belief (Wilson 2002; Alcorta and Sosis 2005). In the absence of such input, genetic predispositions remain latent. Such environmentally cued gene expression permits broad adaptive lability while ensuring optimal allocation of limited resources. The development of religious beliefs and behaviors is likely to reflect such interactive, ontogenetic processes.

While we obviously cannot review all of the burgeoning evolutionary literature on religion, we encourage interested readers to see books by Atran (2002), Boyer (2001), and Wilson (2002), review articles by Bulbulia (2004a) and Sosis and Alcorta (2003), and the wide literature cited therein. Here we focus on recent theoretical developments that synthesize much of the existing literature. Alcorta and Sosis (2005) propose that religion may best be understood as an evolved complex of traits incorporating cognitive, affective, behavioral, and developmental elements. Central to this complex are four cross-culturally recurrent features of religion:

- Communal participation in costly ritual
- Belief in supernatural agents and counterintuitive concepts
- Separation of the sacred and the profane
- Adolescence as the critical life phase for the transmission of religious beliefs and values

Alcorta and Sosis (2005) argue that these traits derive from prehuman ritual systems and were selected for in early hominid populations because they contributed to the ability of individuals to overcome ever-present ecological challenges. By fostering cooperation and extending the communication and coordination of social relations across time and space, these traits served to maximize the potential resource base for early human populations, thereby benefiting individual fitness. In contemporary societies, these four elements continue to provide adaptive benefits and generally co-occur within "religion," but not always. Consequently, Western distinctions between the "secular" and the "religious" are often difficult to discern. For example, the adoption of communal rituals and initiation rites by nominally secular terrorist groups, such as the LTTE, and their quasi-deification of Marxist-Leninist ideals, blurs the line between what is secular and what is religious. These groups engage important elements of the above described religious adaptive complex and reap many of the adaptive benefits achieved by religion.

## Communal Participation in Costly Ritual

Among the most significant challenges terrorists face is ensuring that fellow insurgents are trustworthy and will not defect from the cause (Berman 2003). How can a prospective terrorist guarantee that he will not reveal the locations of hidden conspirators or the secret codes of communication, and that he will not turn aside when asked to carry out a risky or suicidal attack? At first glance, evolutionary theories of religion would appear to hold little promise for answering these questions, or understanding terrorism at all. Natural selection favors genes that get themselves into the next generation, yet terrorists often take great risks with their lives and some of course intentionally sacrifice themselves for their ideological beliefs. Such actions seem to contradict evolutionary expectations. The solution to this puzzle lies in understanding religion as an evolved system of communication, which offers mechanisms that can promote in-group trust and overcome commitment problems (Rappaport 1999; Irons 2001; Atran 2002; Sosis 2003; Atran and Norenzayan 2004; Bulbulia 2004b; Alcorta and Sosis 2005). Irons (2001), for example, posits that the primary adaptive benefit of religion is its ability to foster cooperation and overcome problems of collective action that humans have faced throughout their evolutionary history. The costliness of religious activities, or specifically what Sosis (2006) refers to as the four "Bs"—religious belief, behavior (rituals), badges (such as religious attire), and bans (taboos)—enables them to serve as reliable and honest signals of group commitment. Only those who are committed to the group will be willing to incur the time, energetic, and opportunity costs of religious belief and performance. In other words, adherents pay the costs of religious adherence, but by doing so they demonstrate their commitment and loyalty to the group and can thus achieve a net benefit from successful collective action and other status benefits available to trusted signalers (see Sosis 2003; Sosis and Alcorta 2003; Bulbulia 2004b).

Surprising to many observers, costly religious demands are today increasing in many communities throughout the world. Indeed, the global rise in religious terrorism has been paralleled by a worldwide growth in religious fundamentalism. Fundamentalism typically refers to a religious ideology that embraces scriptural literalism and traditional religious values. Current fundamentalist trends, however, have placed *higher* demands on their practitioners than the traditional practices that they claim to emulate. For example, the standards of *kashrut* (laws pertaining to edible food) among ultraorthodox Jews are more stringent now than at any time in Jewish history (Sosis, in press). Signaling theory suggests three factors that may be motivating the fundamentalist trend toward increasing ritual requirements. First, the rising costs of membership may be a direct response to increases in perceived risk of apostasy faced by religious groups, a risk generated by the rapid improvement in mass media technologies, which expose wide audiences to Western secular values and culture. Second, and somewhat paradoxically, the multicultural openness of Western societies may also contribute to fundamentalist trends. While the celebration of multiculturalism has yet to embrace aggressive fundamentalism, in societies where group differences are tolerated and even encouraged, maintenance of in-group cohesion requires that groups increase their distinctiveness in order to preserve the relative costliness of the group's previous bans and badges. Thus, multiculturalism may actually initiate movements toward fundamentalism, even while vehemently rejecting fundamentalism's message of possessing life's only true path. Notably, Juergensmeyer (2002) observes that one of the universal features of religious terrorists is a strong rejection of Western multiculturalism. Third, signaling theory predicts an increase in signal costs as resource competition escalates. In highly competitive modern multicultural nation-states, the higher costs incurred by religious fundamentalism are likely to be offset by the economic and political gains achievable through religious consolidation and organization of group membership.

The evolutionary signaling theory of religion assumes an inverted-U-shaped relationship between the costliness of religious activity and in-group cooperation. Since imposing costly requirements upon group members is challenging, and greater than optimum costs are expected to negatively impact group cohesion, most groups are predicted to impose less than their optimal level of costly requirements, and thus be observed on the increasing side of the U-shaped distribution. Experimental, cross-cultural, and ethnohistorical research evaluating this prediction has been largely supportive (Sosis 2000; Sosis and Bressler 2003; Sosis and Ruffle 2003, 2004; Ruffle and Sosis 2007; Sosis et al. 2007). Religious terrorists of course employ religiously defined costly requirements to signal commitment, resulting in high levels of in-group cohesiveness and trust that are essential for carrying out their clandestine activities (e.g., Hassan 2001). As Pape describes, terrorists have "a close bond of loyalty to comrades and devotion to leaders; and they have a system of initiation and rituals signifying an individual's level of commitment to the community" (2005, 8). Interestingly, among many terrorist cells these rituals also include the recording of a video testament prior to an attack (Atran 2003). Such video testaments not only serve to immortalize the suicide terrorist and his cause among followers; they also create undeniable contracts, as well. Defecting on a mission after declaring and documenting one's intentions would result in severe psychological, social, and presumably spiritual costs.

Evolutionary signaling theory assumes that the short-term costs of displaying a signal are repaid through individual gains. This creates a particular challenge for understanding suicide terrorism from a signaling theory perspective, since individuals are obviously not around to reap any benefits from their actions. How can suicide terrorism possibly constitute an individually adaptive response? There would appear to be three noncompeting alternative explanations. First, while the individual faces the ultimate sacrifice, suicide terrorism is likely to benefit the group, and Pape's (2003, 2005) analysis showing that groups deploying suicide terrorists tend to achieve their goals would support this interpretation. Suicide terrorism may offer the most promising example of strong selective pressures operating at the group level (Villarreal, this volume). Second, it is possible that suicide bombers recoup their losses through benefits to their kin. For example, the families of Palestinian suicide terrorists receive financial payments (up to $10,000) for their martyred sons and daughters. However, Israel's policy of destroying suicide bomber's homes would appear to counterbalance these indirect fitness gains and be a strong negative incentive to sacrifice oneself for one's family. Noting Krueger and Maleckova's (2002) finding that Hezbollah suicide bombers attained above average education, Azam (2005) argues that suicide bombers may be investing in future kin generations (their higher educations makes them appreciate the importance of investing in the future). However, this poses a significant collective action problem, and it would appear that under most conditions one would be better off letting someone else make the investment (i.e., sacrifice one's life for future generations). A third possibility is that the payoffs motivating suicide bombers are not material but rather otherworldly. Indeed, when applying evolutionary signaling theory to religious activity, both Sosis (2003) and Bulbulia (2004b) incorporate *perceived* gains into their models, which include payoffs attained in the afterlife. They independently found that afterlife payoffs can dramatically alter the dynamic of the game and favor costly religious activity. Moreover, we suspect that not only do martyrs expect to reap their heavenly rewards, but that they also include the reputational benefits they expect to receive as a martyr into their calculations (e.g., Richardson 2006, 124), even though they will of course not be around to enjoy their newly attained status. If afterlife rewards and concerns of postmortem reputation are motivating suicide bombers, such beliefs are likely to be maladaptive, unless kin significantly benefit from being related to a martyr.

## Belief in Supernatural Agents and Counterintuitive Concepts

The second feature of the adaptive religious complex that Alcorta and Sosis (2005) describe concerns supernatural agents and counterintuitive concepts. Evolutionary cognitive scientists have shown that the supernatural agents of religious belief systems are "full access strategic agents" (Boyer 2001). They are "envisioned as possessing knowledge of socially strategic information, having unlimited perceptual access to socially maligned behaviors that occur in private and therefore outside the perceptual boundaries of everyday human agents" (Bering 2005, 418). Furthermore, accumulating research indicates that humans exhibit a developmental predisposition to believe in such socially omniscient supernatural agents, appearing in early childhood and diminishing in adulthood. Cross-cultural studies conducted with children between the ages of 3 and 12 indicate that young children possess an "intuitive theism" that differentiates the social omniscience of supernatural agents from the fallible knowledge of natural social agents, such as parents (Kelemen 2004). By late childhood supernatural agents are not only socially omniscient; they are regarded as agents capable of using such knowledge to reward and punish deeds that are now viewed within a moral framework. Several evolutionary researchers have emphasized the role that supernatural punishment plays in promoting community-defined moral behavior, and specifically in-group cooperation (Johnson et al. 2003; Bulbulia 2004b; Johnson and Kruger 2004; Johnson 2005; Sosis 2005). Recent experimental evidence indicates that "even subtle unconscious exposure to religious ideas can dramatically encourage prosocial over selfish behavior" in theists and atheists alike (Shariff and Norenzayan, 2007).

Evolutionary cognitive scientists have further noted that the counterintuitive concepts that characterize religious beliefs, such as bleeding statues and virgin births, are both attention arresting and memorable (Boyer 2001; Atran 2002). These features make them particularly effective for both vertical (across generations) and horizontal (within generations) transmission and can help explain why religious ideologies, including those of terrorists, often spread quickly through populations. In addition to their mnemonic efficacy, they comprise almost unbreakable "codes" for the uninitiated. Counterintuitive concepts are not readily generated on the basis of intuitive concepts, thus the chances of spontaneously recreating a preexistent counterintuitive concept are exceedingly low. By incorporating counterintuitive concepts within belief systems, religion creates reliable costly signals that are difficult to "fake." Sosis (2003) has argued that repeated ritual performance fosters and internalizes these counterintuitive beliefs, which typically include a nonmaterial system of reward and punishment, including expectations about afterlife activities.

The promise that 72 virgins await a *shahid* is often joked about, but afterlife rewards are a critical feature of successful ideologies that enable terrorist organizations to motivate recruits to carry out their missions. As a Hamas member describes, "We focus his attention on Paradise, on being in the presence of Allah, on meeting the Prophet Muhammad, on interceding for his loved ones so that they, too, can be saved from the agonies of Hell, on the houris [virgins], and on fighting the Israeli occupation and removing it from the Islamic trust that is Palestine" (Hassan 2001, 39). Female martyrs are promised to be the chief of the virgins and exceed their beauty (Richardson 2006, 122). Even kamikaze pilots were assured that they would be "transcending life and death" (Atran 2003, 1535). Detailed experiments by Bering (2006) demonstrate that humans have a natural inclination to believe that some element, typically a soul, survives death. Indeed, most of us, including atheists, have difficulty conceiving of a complete cessation of mental and social activity following death. Nobody knows what it is like to be dead so people attribute to dead agents the mental traits that they cannot imagine being without. Religious and other cultural beliefs serve to enrich or degrade beliefs in the afterlife, but Bering's work suggests that appeals to rational arguments about the irrationality of afterlife beliefs are likely to face strong resistance. As we discuss below, if it is strategically important to alter terrorists' beliefs about the afterlife, the greatest success can be achieved by exposing children and adolescents to alternative belief schemas *before* they are exposed to the afterlife rewards promised by terrorists.

## Separation of the Sacred and the Profane

The separation of the sacred and profane and the emotional power of sanctified symbols are critical for understanding how terrorists utilize religion for their benefit. Religious ritual is universally used to define the sacred and to separate it from the profane (Eliade 1959; Durkheim 1995). As noted by Rappaport (1999), ritual does not merely identify that which is sacred; it *creates* the sacred. Holy water is not simply water that has been discovered to be holy, or water that has been rationally demonstrated to have special qualities. It is, rather, water that has been *transformed* through ritual. For adherents who have participated in sanctifying rituals, the cognitive schema associated with that which has been sanctified differs from that of the profane. Of greater importance from a behavioral perspective, the emotional significance of holy and profane water is quite distinct. Not only is it inappropriate to treat holy water as one treats profane water; it is emotionally repugnant. While sacred and profane things are cognitively distinguished by adherents, the critical distinction between the sacred and the profane is the emotional charging associated with sacred things (Alcorta and Sosis 2005).

It is the emotional significance of the sacred that underlies "faith," and it is ritual participation that invests the sacred with emotional meaning. Extensive research indicates that emotions constitute evolved adaptations that weight decisions and influence actions (Damasio 1994). The ability of religious ritual to elicit both positive and negative emotional responses in participants provides the substrate for the creation of motivational communal symbols. Through processes of incentive learning, as well as classical and contextual conditioning, the objects, places, and beliefs of religious ritual are invested with emotional significance. The use of communal ritual to invest previously neutral stimuli with deep emotional significance creates a shared symbolic system that subsequently weights individual choices and motivates behavior.

It is noteworthy that the sacred may most commonly be encountered as physical space (Eliade 1959). Pape (2005) argues that at the root of each suicide terror campaign is a dispute over land; an occupying power that must be removed from the homeland. Such conditions are ripe for religious symbolism, and indeed, homelands in these conflicts are always publicly perceived as sacred. Pape (2005, 85) comments that "[a]lthough boundaries may be ambiguous and history may be contested, the homeland is imbued with memories, meanings, and emotions." Religious rituals sustain memories, shape meanings, and foster these emotions. Religion's reliance on such emotionally evocative symbols also explains why religious terrorist groups are more successful than secular ones in mobilizing their forces (Bloom 2005). Religious terrorists do not appeal to rational political arguments to win public approval; they rely on sacred symbols imbued with emotional power to enlist followers in their cause.

Once recruits are secured, group solidarity can be further enhanced through negative affect rituals. Neuropsychological research has shown that negatively valenced stimuli are both more memorable and have greater motivational power than positive stimuli (Cacioppo et al. 2002). As a result of this negativity bias, negatively valenced elements of religion provide a more reliable emotionally anchored mechanism for the subordination of immediate individual interests to cooperative group goals (Alcorta and Sosis 2005). Research on the rituals that terrorist cells employ is scant, but apparently deprivation, such as lengthy fasts, is not uncommon (Friedland 1992; Hassan 2001).

## Adolescence as the Critical Life Phase for the Transmission of Religious Beliefs and Values

Adolescent rites of passage comprise one of the most consistent features of religions across cultures. Through rites of passage initiates learn what things constitute the sacred. Rites of passage purposefully engage unconscious emotional processes, as well as conscious cognitive mechanisms. The conditioned association of such emotions as fear and awe with symbolic cognitive schema achieved through these rites results in the sanctification of those symbols, whether places, artifacts, or beliefs. Because such symbols are deeply associated with emotions engendered through ritual, they take on motivational force. When such rites are simultaneously experienced by groups of individuals, the conditioned association of evoked emotions with specific cognitive schema creates a cultural community bound in motivation, as well as belief.

The human brain demonstrates great plasticity during development. Infancy, childhood, adolescence, and adulthood are marked by differentiated growth patterns in various brain cortices and nuclei (Alcorta 2006). The differential patterns of brain growth across the life course create sensitive periods for particular types of learning. The unique changes occurring in the adolescent brain render this a particularly sensitive developmental period in relation to social, emotional, and symbolic stimuli, precisely the types of stimuli of greatest importance in adolescent rites of passage. Adolescence is also marked by heightened emotional response, and a shifting of the brain's reward circuitry. Adolescence thus comprises a critical period for the learning of emotionally valenced symbolic systems and for integrating these systems into the brain's reward circuitry. Increased risk taking during adolescence affords additional inputs into this system (Alcorta and Sosis 2005). It is therefore not surprising that most terrorists begin their militant life during adolescence. Victoroff (2005, 28) suggests that the "typical development of terrorist sympathies perhaps follows an arc: young adolescents are plastic in their political orientation and open to indoctrination. Positions harden later in adolescence, . . .[and] many retired 'terrorists' reveal a mellowing of attitude." Of course, by the time those raised in a culture of martyrdom reach adolescence they are already prepared to sacrifice themselves without further indoctrination (Brooks 2002; Atran 2003). Bloom observes that by age six Palestinian boys and girls report that they wish to grow up and become *isitshhadis* (martyrs). "By the age of 12, they are fully committed and appreciate what becoming a martyr entails" (2005, 88). As a senior member of the Palestinian group al-Qassam declares, "it is easy to sweep the streets for boys who want to do a martyrdom operation" (Hassan 2001, 39). Nonetheless, the profile of those who actually carry out suicide attacks may be somewhat older (Hassan [2001] reports a range of 18 to 38 among Palestinians), suggesting that the enthusiasm of youth must be balanced with training and the development of trust to carry out such a mission.

## Secular Terrorism and the Evolution of Religion

One advantage of the evolutionary approach we offer here is that by delineating the core adaptive features of religion that facilitate cooperation we can avoid definitional quagmires concerning what constitutes religion. This is important because we suspect that similar to their religious counterparts, successful secular terrorists employ some of these core features, such as emotionally evocative symbols, rituals, and myths. For example, although the Tamil Tigers are not religious, they in fact use "Hindu symbols for purposes of recruitment . . . and rely on the language of religious martyrdom to justify and reward the sacrifice" (Berman and Laitin 2005, 28). And similar to the function of video testaments, prior to suicide missions Tamil Tigers partake in a ritual dinner with their leader, obviously sealing their commitment to carry out the attack (Gambetta 2005). Furthermore, thousands attend the yearly Heroes' Day ritual celebration held on July 5, commemorating the LTTE martyrs, whose shrines are adorned with flowers and trees (Pape 2005, 193). Therefore, the secular-religious distinction made by Western societies with institutionalized religious systems may not be a useful paradigm for examining the determinants of terrorist activity. Rather, analyses would be better served by concentrating on how terrorist organizations use the particular characteristics of the human religious adaptive complex we have outlined here to inspire group commitment and individual action.

## Future Areas of Research

Our brief review of the emerging evolutionary literature on religion and its relation to terrorism leaves many questions unanswered and points to the need for considerable research. We consider a few areas that we believe deserve particular attention.

1. Various researchers note the vital role that charismatic leaders play in shaping recruits' opinions and motivations. For example, Atran (2003, 1537) argues that terrorist leaders manipulate "emotionally driven commitments" of their recruits for the "benefit of the organization rather than the individual." Evolutionary theories of religion have yet to fully address the role of leadership in coordinating behavior within religious groups, but recent evolutionary work on the dynamics of leaders and followers should provide valuable insights (Van Vugt 2006).

2. Argo (2003) contends that in martyrdom cultures individual risk is positively correlated with status benefits. We suggested that martyrs include in their cost-benefit assessments the status benefits they expect to receive following their death, even though they will of course not be around to enjoy their newly attained status. This needs to be studied, as it is likely to be important for understanding some of the benefits that martyrs believe their actions will produce, and offer clues into how those perceived payoffs can be altered.

3. Ritual is a vital mechanism that terrorists employ to secure commitment, but detailed studies on the ritual lives of terrorists are limited. Further-

more, our analyses suggest that adolescence is the critical development phase during which terrorists are created, and thus adolescents should be the focus of considerable terror-related research.

4. Selectionist logic suggests that high-risk behaviors, such as terrorism, are more common in high-fertility populations (e.g., Daly and Wilson 1988). We are not aware of any research that links terrorism, religiosity, and fertility decisions, but this would be a worthwhile pursuit. If there is a positive relationship between terror and fertility, encouraging demographic transitions through such means as expanded female educational and economic opportunities may be one means of reducing terror activity.

5. While adherents generally view religious dogma as unchanging and even eternal, religions are in fact flexible and they do adapt to socioecological circumstances (Rappaport 1999). Rigorous studies that examine the determinants of religious change, especially among extremist forms of religion, are urgently needed.

## Policy Implications

A number of policy implications previously recommended by other researchers are consistent with our evolutionary analysis.

1. As noted above, terrorists have been successful in bringing the world to view their conflicts through a cosmic lens. While it is tempting for governments to rely on absolute moral dichotomies in order to rally their own public in support of whatever military actions they must take, this serves to endorse the terrorists' position and facilitates terrorist recruitment efforts. As Juergensmeyer recommends (2004c), and we concur, governments should avoid depicting conflicts as a cosmic clash between civilizations. Because of the emotional strength of religious symbols, religion is a potent mobilizer of public support, but it is not a vital tool for Western societies and may, conversely, be detrimental given the multicultural constituency of Western societies. Framing conflicts as cosmic clashes is far more valuable for terrorist organizations. Any gains that Western governments achieve by casting the conflict in religious terms will be offset by gains terrorists receive from framing the conflict in this way.

2. Berman and Iannaconne (2005) argue that fostering an open market of religious expression will reduce religious violence. We agree. Trying to eliminate religion, as some recent authors have proposed, is pointless. Evolutionary work not only affirms the empirical resilience of religion, it offers an explanation for why religion endures even in the face of persecution and indicates that the components of religion are highly effective human adaptations. Terrorists are ideologically motivated, which

means that to effectively fight terrorism, it is necessary to compete ideologically as well as militarily. Our analysis highlights the features that make some ideologies more successful than others. While a diversity of ideas can undermine extremist positions, it is critical that new ideas not be perceived as foreign, and that the symbols used to convey these ideas be consistent with existent symbolic systems. As we note, religious fundamentalists believe they are returning to *their* tradition and will reject anything they perceive as not stemming from their past. Moreover, liberal religious ideologies are not compelling substitutes for those seeking an all-encompassing religious life. The goal is not to eradicate fundamentalism, which will not succeed; rather, the policy strategy should be to encourage an open market of religious expression in which nonviolent forms of religious fundamentalism can successfully compete, as can be observed throughout North America.

3. While we agree with much of Atran's (2003) critique of rational choice explanations of suicide terrorism, there is actually considerable overlap between evolutionary and rational choice approaches. The club goods model offered by Berman (2003), Berman and Laitin (2005), and Berman and Iannaccone (2005) is similar to the evolutionary signaling model of religion (relying on rational choice rather than evolutionary assumptions) and is in fact more fully developed and more rigorously tested. Here we concur with Berman and colleagues that it is critical to recognize religious requirements as commitment signals, since governmental policies can unintentionally undermine these signals, forcing fundamentalist groups to generate even costlier signals (e.g., Berman 2000). The primary difference between the rational choice and evolutionary approach is that evolutionists, such as Atran and ourselves, maintain that emotions are vital for making sense of terrorism. Recent neuropsychological research underscores the critical, though frequently unconscious, role of emotions in individual decision making (Cardinal et al. 2002). We view the ability of terrorist groups to shape the emotions of their supporters and recruits as a critical element in their success. As Victoroff notes, "passion often trumps rationality" (2005, 17).

Evolutionary research on religion also suggests several policy recommendations we have not previously seen in the literature.

1. Our analysis indicates that religious emotions run deep. Therefore, once conflicts have been framed in religious terms, religious groups must be represented in negotiations. We echo Bloom's warning that religious groups "are more complicated and dangerous negotiating partners" (2005, 98). However, without their participation and endorsement of a solution, negotiated peace between secular groups will not satisfy aggrieved parties and will not endure.

2. Signaling theory suggests that in-group members who live in communities that demand costly badges of identity will be best able to detect terrorists posing as group members. For example, Hoffman (2003) describes Palestinian suicide bombers who dress as ultraorthodox Jews in order to penetrate deep into civilian areas. We recommend training ultraorthodox security guards to assist Israeli soldiers already posted at strategic locations (bus stations, market entrances, etc.).

3. We agree with Atran's (2003) conclusion that the best defense against terrorism is to prevent potential recruits from joining terrorist organizations. Our analysis further points to the importance of adolescence as the critical time in which individuals will be receptive to such groups. To prevent adolescents from joining terrorist organizations, alternative adolescent youth activities must be developed, funded, and encouraged. We view this as essential.

4. The ideological battle cannot be waged only among adults. Our analysis emphasizes the importance of childhood exposure to religious ideas and concepts and the importance of early ritual activity in fostering long-term belief. Alternative forms of religious expression that are publicly perceived as traditional must be fostered and made available to children and adolescents. This will mean developing youth activities as suggested above, and also funding attractive schools that can draw students away from terrorist funded schools and provide them with both ideologies and opportunities for a nonterrorist future.

5. From an evolutionary perspective, religion can be seen as an adaptive proximate mechanism for creating cohesive, cooperative groups. Modern nation-states with educational, economic, and political policies that equitably integrate their multicultural constituencies and effectively sanction policy transgressions reduce the competitive need for religious fundamentalism and increase the opportunity costs such fundamentalism incurs. Building the economic resource base of developing nations, and creating equitable and effective political structures to insure equal access to that base are fundamental steps in eradicating religious terrorism throughout the world.

## Conclusion

Evolutionary studies of religion are critical for understanding the current world of terror. The novel insights offered by evolutionary theories of religion are not limited to religiously motivated terrorism, but they do explain why religiously motivated terrorism is on the rise and is particularly lethal. While there is considerable overlap between evolutionary and rational choice approaches to terror, we believe that evolutionary approaches may fill in significant gaps that have remained unexplained by the dominant rational choice paradigm. The reach of evolutionary logic has extended to studies in history, literature, art, and recently religion; the time is ripe to apply the insights of evolutionary theory to terrorism as well.

ACKNOWLEDGMENTS

We thank Dominic Johnson, Bradley Ruffle, and Rafe Sagarin for valuable comments on earlier drafts of this manuscript.

REFERENCES

Alcorta, C. 2006. Youth, religion, and resilience. Doctoral dissertation. Storrs: University of Connecticut.

Alcorta, C., and R. Sosis 2005. Ritual, emotion, and sacred symbols: The evolution of religion as an adaptive complex. *Human Nature* 16: 323–359.

Argo, N. 2003. The banality of evil: Understanding today's human bombs. Policy paper, Preventative Defense Project. Palo Alto: Stanford University.

Atran, S. 2002. *In gods we trust: The evolutionary landscape of religion.* Oxford: Oxford University Press.

Atran, S. 2003. Genesis of suicide terrorism. *Science* 299: 1534–1539.

Atran, S., and A. Norenzayan. 2004. Religion's evolutionary landscape: Counterintuition, commitment, compassion, communion. *Behavioral and Brain Sciences* 27: 713–730.

Azam, J.-P. 2005. Suicide-bombing as intergenerational investment. *Public Choice* 122: 177–198.

Bering, J. 2005. The evolutionary history of an illusion: Religious causal beliefs in children and adults. In *Origins of the social mind: Evolutionary psychology and child development,* ed. B. Ellis and D. Bjorklund, 411–437. New York: Guilford Press.

Bering, J. 2006. The cognitive psychology of belief in the supernatural. *American Scientist* 94: 142–149.

Bering, J. 2006. The folk psychology of souls. *Behavioral and Brain Sciences* 29: 453–462.

Berman, E. 2000. Sect, subsidy and sacrifice: An economist's view of ultraorthodox Jews. *Quarterly Journal of Economics* 115: 905–953.

Berman, E. 2003. Hamas, Taliban, and the Jewish underground: An economist's view of radical religious militias. NBER Working Paper 10004. National Bureau of Economic Research, Cambridge, MA.

Berman, E., and L. Iannaconne. 2005. Religious extremism: The good, the bad, and the deadly. NBER Working Paper 11663. National Bureau of Economic Research, Cambridge, MA.

Berman, E., and D. Laitin. 2005. Hard targets: Theory and evidence on suicide attacks. NBER Working Paper 11740. Cambridge, MA: National Bureau of Economic Research.

Berrebi, C. 2003. Evidence about the link between education, poverty, and terrorism among Palestinians. Princeton University Industrial Relations Section Working Paper 477. Princeton, NJ: Department of Economics, Princeton University.

Bloom, M. 2005. *Dying to kill: The global phenomenon of suicide terror.* New York: Columbia University Press.

Boyer, P. 2001. *Religion explained: The evolutionary origins of religious thought.* New York: Basic Books.

Brooks, D. 2002. The culture of martyrdom. *Atlantic Monthly* 289: 18–20.

Bulbulia, J. 2004a. Religious costs as adaptations that signal altruistic intention. *Evolution and Cognition* 10: 19–38.

Bulbulia, J. 2004b. Area review: The cognitive and evolutionary psychology of religion. *Biology and Philosophy* 18: 655–686.

Cacioppo, J. T., W. L. Gardner, and G. G. Berntson. 2002. The affect system has parallel and integrative processing components: Form follows function. In *Foundations in social neuroscience,* ed. J. T. Cacioppo, G. G. Berntson, R. Adolphs, C. S. Carter, R. J. Davidson, M. K. McClintock, B. S. McEwen, M. J. Meaney, D. L. Schacter, E. M. Sternberg, S. S. Suomi, and S. E. Taylor, 493–522. Cambridge: MIT Press.

Cardinal, R., J. Parkinson, J. Hall, and B. Everitt. 2002. Emotion and motivation: The role of the amygdala, ventral striatum and prefrontal cortex. *Neuroscience and Biobehavioral Reviews* 26: 321–352.

Daly, M., and M. Wilson. 1988. *Homicide.* New York: Aldine de Gruyter.

Damasio, A. 1994. *Descartes' error: Emotion, reason, and the human brain.* New York: Avon Books.

Durkheim, E. 1995. *The elementary forms of the religious life.* New York: The Free Press.

Eliade, M. 1959. *The sacred and the profane: The nature of religion.* New York: Harcourt Brace Jovanovich.

Friedland, N. 1992. Becoming a terrorist: social and individual antecedents. In *Terrorism: Roots, impact, responses,* ed. L. Howard, 81–93. New York: Praeger.

Gambetta, D. 2005. *Making sense of suicide missions.* Oxford: Oxford University Press.

Hassan, N. 2001. An arsenal of believers: Talking to the "human bombs." *The New Yorker.* 19 (November): 36–41.

Hoffman, B. 1998. *Inside terrorism.* New York: Columbia University Press.

Hoffman, B. 2003. The logic of suicide terrorism. *The Atlantic Monthly* 291: 40–47.

Irons, W. 2001. Religion as a hard-to-fake sign of commitment. In *Evolution and the capacity for commitment,* ed. R. Nesse, 292–309. New York: Russell Sage Foundation.

Johnson, D. 2005. God's punishment and public goods: A test of the supernatural punishment hypothesis in 186 world cultures. *Human Nature* 16: 410–446.

Johnson, D., and O. Kruger. 2004. The good of wrath: Supernatural punishment and the evolution of cooperation. *Political Theology* 5: 159–176.

Johnson, D., P. Stopka, and S. Knights. 2003. The puzzle of human cooperation. *Nature* 421: 911–912.

Juergensmeyer, M. 2002. Religious terror and global war. Global and International Studies Program, Paper 2. Santa Barbara: University of California. http://repositories.cdlib.org/gis/2.

Juergensmeyer, M. 2003. *Terror in the mind of God: The global rise in religious violence.* Berkeley: University of California Press.

Juergensmeyer, M. 2004a. From Bhindranwale to Bin Laden: The rise of religious violence. Presentation at Arizona State University, Tempe, October 14–15.

Juergensmeyer, M. 2004b. Holy orders: Opposition to modern states. *Harvard International Review* 25: 34–38.

Juergensmeyer, M. 2004c. Is religion the problem? *Hedgehog Review* 6: 21–33.

Kelemen, D. 2004. Are children "intuitive theists"? Reasoning about purpose and design in nature. *Psychological Science* 15: 295–301.

Krueger, A., and J. Maleckova. 2002. Education, poverty, political violence and terrorism: Is there a causal connection? NBER Working Paper 9074. Cambridge, MA: National Bureau of Economic Research.

Lincoln, B. 2003. *Holy terrors: Thinking about religion after September 11.* Chicago: University of Chicago Press.

Pape, R. 2003. The strategic logic of suicide terrorism. *American Political Science Review.* 97: 343–361.

Pape, R. 2005. *Dying to win: The strategic logic of suicide terrorism.* New York: Random House.

Rappaport, R. 1999. *Ritual and religion in the making of humanity.* Cambridge: Cambridge University Press.

Richardson, L. 2006. *What terrorists want.* New York: Random House.

Ruffle, B., and R. Sosis. 2007. Does it pay to pray? Costly rituals and cooperation. *The BE Press of Economic Policy and Analysis (Contributions)* 7:1–35 (Article 18).

Sageman, M. 2004. *Understanding terror networks.* Philadelphia: University of Pennsylvania Press.

Shariff, A., and A. Norenzayan. God is watching you: Supernatural agent concepts increase prosocial behavior in an anonymous economic game. *Psychological Science* 18:803–809.

Simon, S., and D. Benjamin. 2000. America and the new terrorism. *Survival* 42: 59–75.

Sosis, R. 2000. Religion and intragroup cooperation: Preliminary results of a comparative analysis of utopian communities. *Cross-Cultural Research* 34: 70–87.

Sosis, R. 2003. Why aren't we all Hutterites? Costly signaling theory and religion. *Human Nature* 14: 91–127.

Sosis, R. 2005. Does religion promote trust? The role of signaling, reputation, and punishment. *Interdisciplinary Journal of Research on Religion* 1: 1–30 (Article 7).

Sosis, R. 2006. Religious behaviors, badges, and bans: Signaling theory and the evolution of religion. In *Where God and science meet: How brain and evolutionary studies alter our understanding of religion,* vol. 1: *Evolution, genes, and the religious brain,* ed. Patrick McNamara, 61–86. Westport, CT: Praeger Publishers.

Sosis, R. In press. Why are synagogue services so long? An evolutionary examination of Jewish ritual signals. In *Judaism and bio-psychology,* ed. Rick Goldberg.

Sosis, R., and C. Alcorta. 2003. Signaling, solidarity and the sacred: The evolution of religious behavior. *Evolutionary Anthropology* 12: 264–274.

Sosis, R., and E. Bressler. 2003. Cooperation and commune longevity: A test of the costly signaling theory of religion. *Cross-Cultural Research* 37: 211–239.

Sosis, R., and B. Ruffle. 2003. Religious ritual and cooperation: Testing for a relationship on Israeli religious and secular kibbutzim. *Current Anthropology* 44: 713–722.

Sosis, R., and B. Ruffle. 2004. Ideology, religion, and the evolution of cooperation: Field tests on Israeli kibbutzim. *Research in Economic Anthropology* 23: 89–117.

Sosis, R., H. Kress, and J. Boster. 2007. Scars for war: Evaluating alternative signaling explanations for cross-cultural variance in ritual costs. *Evolution and Human Behavior.* 28:234–247

Van Vugt, M. 2006. Evolutionary origins of leadership and followership. *Personality and Social Psychology Review* 10: 354–371.

Victoroff, J. 2005. The mind of the terrorist: A review and critique of psychological approaches. *Journal of Conflict Resolution* 49: 3–42.

Wilson, D. 2002. *Darwin's cathedral: Evolution, religion, and the nature of society.* Chicago: University of Chicago Press.

Zuckerman, P. 2006. Atheism: Contemporary rates and patterns. In *Cambridge companion to atheism,* ed. M. Martin. Cambridge: Cambridge University Press.

Chapter 8

# CAUSES OF AND SOLUTIONS TO ISLAMIC FUNDAMENTALIST TERRORISM

BRADLEY A. THAYER

Long before the terror attacks of 9/11, the United States faced the threat of Islamic fundamentalist terrorism. Throughout the decade before the horrible attacks of that day, the United States had been repeatedly attacked but had chosen not to address the threat directly or effectually. Post-9/11, the counterterrorism policy of the United States is the antipode of what it was. Campaigns in Afghanistan, Pakistan, Iraq, the Horn of Africa, the Philippines, Indonesia, the Sahel, Maghreb, Thailand, Malaysia, and in European states have done much to weaken the most notorious group, al Qaeda.[1] That is progress to be sure. But the jury is still out on how great a threat this form of terrorism remains for the United States.

Fundamentally, this is because its opponents do not know how to measure its health—is it a well nourished organism? Is it able to reproduce? It is still able to conduct attacks, as 10/12 in Bali, and 3/11 in Spain demonstrate? Indeed, more than 800 people have been killed in over 14 attacks blamed on al Qaeda since 9/11, including attacks that have occurred in Bangladesh, Egypt, Great Britain, Kenya, Pakistan, Turkey, Tunisia, regularly in Iraq, and elsewhere. Of course, these attacks are not of the type as before 9/11 and indicate that the organization or related organizations are conducting spasmodic and very traditional forms of attack—principally bombings—which have been used by terrorists since the nineteenth century. Additionally, the January 2006 offer by Usama bin Laden of a truce to the United States suggests the organization is weakened. However, even if al Qaeda is weakened, it is not the totality of Islamic fundamentalist terrorism, which remains a danger.

Islamic fundamentalist terror is well studied through the lens of international politics, economics, and cultural studies. Each of these approaches offers insights into the motivation and recruitment policies of these groups. This brief chapter approaches the topic differently. My central arguments are that the application of concepts and approaches from the life sciences yields new insights into, first, the deep causes of Islamic fundamentalist terrorism; second, the motivation of the terrorists; and third, solutions to this form of terrorism. A consilient approach, incorporating ideas from the life as well as social sciences, will aid the social scientists and policy analysts who have traditionally studied the problem of Islamic fundamentalist terrorism because it provides direct policy applications to identify its causes and solutions.[2] In addition, as other scholars in this book and elsewhere have demonstrated, evolution or ecological ideas may be used as a metaphor, analogy, or heuristic, to allow scholars and analysts to approach the problem in new ways and to derive new solutions (Sprinkle 2006).

My argument is important for three reasons. First, understanding the motivations of terrorists is critical for creating efficient policies to stop them, ideally before they ever become terrorists. Second, this approach allows policymakers to understand why few Islamic fundamentalist terrorists defect and how policies may be created to promote defections. Finally, this chapter has intellectual significance, as does this book, because it advances the goal of consilience. That is, the authors use insights from human evolution and ecology, as well as from the social sciences, to create a more comprehensive and detailed understanding of human behavior. In essence, consilient approaches bridge the gap between the life and social sciences. For the advancement of knowledge concerning human behavior, there is no more important work than trying to remove the barriers between the life and social sciences to promote mutual understanding and interaction.

Indeed, Islamic fundamentalist terrorism must be studied through a consilient approach due to the magnitude of the threat it poses but also because the life sciences have much to offer the study of terrorism.[3] Scholars from all disciplines need to unite to conduct the equivalent of a human genome project for Islamic fundamentalist terrorism—to understand the motivation of terror groups and why they maintain the fealty of their members so that policies may be created to defeat them. If we do not study this issue as intensely as the human genome, these terrorists may kill us. Westerners have discovered since 9/11, 10/12 (Bali), 3/11 (Madrid), and 7/7 (London) that they might not be interested in Islamic fundamentalist terrorism, but the terrorists are interested in them.

My analysis examines some of the causes of Islamic fundamentalist terrorism and offers calculated solutions to it. To accomplish this, first, I postulate its deep causes. These are the conditions that have allowed this form of terrorism to develop, rather than the proximate causes, for example, what would make someone decide to become a terrorist, or what tactics a terrorist organization would use to maintain the loyalty of its members. My use of the two terms, however, allows me to acknowledge the fundamental difference between the immediate causes of terrorism and the fundamental or permissive causes of terrorism.[4] There are three ultimate causes: anarchy in international politics; the dominance, or hegemony, of the United States in international politics; and the nature of Islam. I address each of these in turn before turning to the proximate causes, after which I advance solutions based on deep and proximate causation.

Of course, every analysis has boundaries, and this brief chapter is no exception. I do not attempt to explore all of the causes of Islamic fundamentalist terrorism, why states sponsor terrorism, or the myriad motivations that might lead one to suicide terrorism or to support al Qaeda and allied groups (Roy 2004, 42–54, 247–289; Bloom 2005). Additionally, I do not consider the issue of female suicide terrorists founded in the Chechen struggle, among the Liberation Tigers of Tamil Eelam, and in some Palestinian groups (Khosrokhavar 2005, 217–219). Each of these issues is important to study and can be informed by the life sciences, but I regret they fall outside the narrower scope of the present chapter.

## The Deep Causes of Islamic Fundamentalist Terrorism

International politics is anarchic. This means that there is no governmental authority above the level of the state in international politics. That is the key difference between domestic politics, where there is hierarchic government, and international politics, where there is none to adjudicate disputes and provide protection for citizens (Waltz 1979, 114–116). The anarchic structure of international politics has important results. One is because the international system is anarchic, international politics is a dangerous environment where war or human rights abuses are always possible. The lack of governance also means that failed states are always a possibility. Failed states are countries like Sudan or Afghanistan under the Taliban, where sovereignty is contested by different political factions, and the great powers—the United States, France, or Russia—are content to allow ambiguous sovereignty to remain. The fact that Sudan was a failed state in the 1990s allowed al Qaeda to nest there. Similarly, the failed state of Afghanistan under the Taliban allowed al Qaeda to flourish, train a new generation of terrorists, and become the threat it, its spin-offs, and related groups are today.

The hegemony of the United States is the second deep cause. The end of the Cold War left the United States as the only superpower and the world's military, economic, and ideological leader until the rise of a new peer-competitor like China. The profound power of the United States allows it to use military force and support allied regimes the world over, including many in the Arab world. In 1990, Iraq's attack on Kuwait allowed the United States to base forces in Saudi Arabia and to expand significantly its military forces in the Persian Gulf. Those forces remained at the conclusion of the war. The worldwide presence and power of the United States means that it is an object of terrorism.

A consistent and important rationale used by Islamic fundamentalist terrorists is to remove the presence of the United States in the Islamic World, and Saudi Arabia most importantly. The latter, at least, was achieved shortly before the commencement of Operation Iraqi Freedom, when all permanently stationed United States forces withdrew to locations in the Persian Gulf.

A second rationale for attacks that stems from the power of the United States is its support given to "apostate" regimes in the Arab world. To protect its worldwide interests the United States has established a network of alliances in the Middle East—from Morocco to the Gulf sheikhdoms and from Turkey to the Horn of Africa and Somalia. Islamic fundamentalist terrorists are the sworn enemy of these governments. Indeed, most Islamic fundamentalist terrorism is directed against these governments and their supporters—what the jihadists call the "near enemy"—who are reviled because they are seen as apostates who reject Islam to embrace a secular ideology and ally with the United States (Gerges 2005, 43–79). In turn, the United States is considered the "far enemy" (Hafez 2003, 187–190; Gerges 2005, 119–184).

Until al Qaeda, the United States was rarely attacked by Islamic fundamentalist terrorists; in the 1980s, Hizbullah attacked U.S. forces in Lebanon until they withdrew for nationalistic, not religious, reasons. The rationale for the attacks against the United States' military and allies, and in the United States itself, flows from the terrorists' grand strategic goal of forcing the United States to withdraw its forces from the Middle East and to terminate support for its allies. Al Qaeda's lasting contribution to Islamic fundamentalist terrorism was to target the "far enemy" first, before the apostate regimes. If attacked sufficiently, the United States would withdraw from the region, and perhaps globally. Deprived of United States' support, the apostate regimes will fall, and the Islamic community will be pure.

Not surprisingly, a narrow, fundamentalist interpretation of the religion of Islam is a deep cause as well. Islamic fundamentalist terrorism finds its justification in the ideologies of Salafiyya or Wahhabism, which in turn are anchored in fundamentalist interpretations of the Qur'an and Sunnah (or hadith), the sayings and traditions of the prophet Muhammad.[5] The validity of this interpretation of Islam is disputed by Islamic scholars, but for the terrorists themselves and many of their sympathizers the dispute is irrelevant. A consilient approach permits scholars to understand religion in a new light—an evolutionary explanation of religious beliefs. There is a well-developed literature on this subject, and perhaps the best analysis is provided by David Sloan Wilson's scholarship, which allows scholars to understand how religion meets human needs (2002).

More importantly for the present purposes, consilient scholarship also allows scholars to grasp why the religion of Islam would be attractive for individuals. While some of these features would be shared with other religions, some are unique to Islam. It is quite reasonable to expect that certain behaviors and beliefs would evolve in an anarchic, resource-scarce environment. It is also reasonable to expect that these behaviors might last longer in relatively isolated parts of the Muslim world, such as the Arabian Peninsula, that sought little contact with the outside world, than in more the cosmopolitan parts of the Muslim world such as Algiers, Istanbul, or Jakarta. Comprehending such behaviors aids those countries fighting the War on Terror because some should be nurtured and some must be combated to decrease the attractiveness of Islamic fundamentalist terrorism.

First, like almost all religions, Islam provides order and a sense of hierarchy—an important influence on social animals who evolved in dominance hierarchies as did humans. Second, the dominant religions arose at a time of great resource scarcity—an environment very close to the environment in which humans evolved over the last 4 to 5 million years—the environment of evolutionary adaptation. Resource scarcity has had a profound influence on human evolution and, accordingly, human behaviors. Individuals in those environments would have difficulty accessing potable water, food, productive land, housing in safe, relatively disease-free locales, and access to mates.

The latter is particularly important in the polygamous societies of the Arabian Peninsula and throughout the Middle East. Generally, if one male has multiple wives and male:female birthrates remain approximately equal, there would be fewer available females for the remaining males. Those males without mates would have to resort to extraordinary acts to increase their social status to allow them to attract a mate. Islam, unlike Christianity and Judaism, continues to promote polygamy. The practice is most common in Saudi Arabia, the provenance of 15 of 19 of the 9/11 hijackers. Consequently, low-status males have to resort to extraordinary measures to increase their attractiveness as mates for themselves or their surviving relatives. In sum, for individuals in the ecological environment of the Arabian Peninsula, the promise of a resource abundant environment in the afterlife—if he is good (that is, a true Muslim)—would be attractive and credible, and so would greatly influence behavior in this world.

## The Proximate Causes of Islamic Fundamentalist Terrorism

The life sciences also generate insights at the proximate level of causation—in what people believe motivates them to conduct acts of terrorism and some of the tactics used by the terrorists to maintain support. These are important to recognize in order to generate effective solutions. I now examine the three proximate causes of Islamic fundamentalist terrorism that influence the individual motivation of Islamic fundamentalist terrorists: desire to dominate, feelings of emasculation, and the polygamous marriage practices and related demographics of the Middle East. I then address the reasons why Islamic fundamentalist terrorist organizations inspire such fealty from their members. I argue that this is because they mimic the family.

## Motivation of the Individual

An evolutionary perspective allows us to grasp why individuals would choose to become terrorists. The first reason is a desire to dominate, or to have your will imposed on others. This is such a common cause of human behavior that it is almost taken for granted, although naturally, there is variation in its effect. People commonly speak of having a desire "to make a difference," or to have "influence"; but the trait evinces itself in greater forms typically found among captains of industry who seek to dominate their market, leaders of countries who seek to dominate their domestic and international opponents, athletes who want to dominate the other team or their sport, and some individuals who choose to become terrorists. Evolutionary theory explains this—the desire to dominate is a trait (Thayer 2004, 71–75).

Seen through the lens of evolutionary theory, domination typically means that certain individuals in social groups have regular priority of access to resources in competitive situations (Frank 1985). For most social mammals, including humans, a form of social organization called a dominance hierarchy operates most of the time. The creation of the dominance hierarchy may be violent and is almost always competitive. A single leader, the alpha male, leads the group and controls resources. The ubiquity of this social ordering strongly suggests that such a pattern of organization contributes to fitness. Ethologists argue that dominance hierarchies evolve because they help defend against predators, promote the harvesting of resources, and reduce intragroup conflict (McEachron and Baer 1982; Lopreato 1984; Moyer 1987).[6] Humans, and males in particular, who live in a tight dominance hierarchy will be looking for ways in which they can have an impact or call attention to themselves to increase their social status and attractiveness as mates, as well as to have a feeling of empowerment—of belonging to a dynamic and strong organization—which, as John Tooby has noted, is particularly attractive to males (2005). Joining a terrorist group or supporting the fundamentalism that nourishes the terrorist groups are ways people may feel emboldened or more powerful. They are participants in a socially attractive cause that will translate into social and sexual rewards.

Second, again for males in particular, joining a terrorist group will reduce feelings of emasculation caused by a perception of the weakness of Muslim countries and Arab countries in particular against a powerful West—whether that is the United States, France, or Israel (Khosrokhavar 2005, 130–133). The concept of virility may be hard for a Western academic audience to comprehend given the efforts to erase it in Western societies. However, despite Western, elite opinion, masculinity has traction in other societies, and no less so in Muslim societies. Indeed, Nathaniel Fick, a former Marine Corps platoon leader who fought in Operation Iraqi Freedom in 2003 reports in his account of that war that all social groups welcomed liberation from Saddam Hussein except males in their twenties and thirties, because the Marines were doing the job of the Iraqis, they were doing what the Iraqis could not—ridding the country of the tyrant—and felt emasculated as a consequence (2005).

Given this power imbalance, it is transparent why terrorism is a route some young men choose. After all, terrorism is often called "the weapon of the weak." Evolution allows us to understand why males, and principally adolescent or young males, will be attracted to dynamic ideologies promising to overturn the dominance hierarchy. In sum, what the fundamentalists would like is to have the Western alpha male supplanted by the Islamic alpha male.

Third, much of the Arab world is polygamous. Polygamy means mates for some men and none for others. The implications of this are profound for those without mates and little prospect of gaining one. The path of the jihadi is a solution to this problem (Kanazawa 2005). In fact, many people in the West or outside the Muslim world know little of Islam other than its promise of 72 virgins for a jihadist.[7] Too few are aware of the promise that the jihadist may bring 70 members of his household into Paradise with him (Burke 2003, 69; Stern 2003). Although failed suicide bombers may not admit this objective, perhaps considering it too vulgar or impious, it can be a key draw for a male denied access to females by a polygamous society. In March 2004, Hussam Abdo (also Abdu), a 16-year-old failed suicide bomber who was caught at an Israeli checkpoint in Gaza, explained to Israeli intelligence officials that his dwarfism made him the object of ridicule at school and unlikely to find a girlfriend (Harel 2004). The purported reward of the suicide bomber was thus attractive to him.

Although it would never occur to those who study this problem through purely the social lens, Islamic fundamentalist terrorism may be considered a male mating strategy. This is because it offers males without females, and with little hope of gaining a socially positive marriage, the opportunity to do so due to the social attractiveness of the terrorists in much of the Muslim world, and with the promise of considerable sexual reward for the jihadist (Khosrokhavar 2005, 134).

Of course, this does not contribute to the suicide bomber's actual fitness—obviously, it is the reverse. However, the faith of the suicide bomber is what is salient here, and the reality of the afterlife for the suicide terrorist is real and attractive enough for him to undertake a suicide mission. Moreover, the suicide terrorist will know that he may be venerated in some circles for his suicide action. In turn, this will increase the attractiveness of his siblings as mates. Thus, inclusive fitness allows the suicide terrorist to be understood in a manner not grasped by social explanations of suicide terrorism.

## Why Terrorists Inspire Such Fealty among Their Members

Just as evolution allows us to grasp the motivation of the terrorist, it aids understanding the behavior of terrorist groups as well, particularly why the group inspires such fealty from its members. From publicly available information, Islamic fundamentalist terrorism groups suffer few defections in membership, unlike the ideological terrorist groups of the Cold War such as the Red Army Faction in West Germany or the Red Brigades in Italy, who suffered chronic defections that contributed to their collapse. This is the case not only of al Qaeda but of most of the Islamic fundamentalist terror groups.

The major reason why is because the terrorist organization mimics the family. It is no accident that the terrorists refer to one another as "brothers." Evolutionary psychology explains why that will be effective. Humans evolved in small groups—principally the family and extended-family hunter groups. Accordingly, the human mind is well suited for comprehending and bonding with small groups of dozens, or, at most, 100 or 150 people (Tooby 2005). To be sure, they may bond in larger units, for example a country, but that requires an extensive effort by the state, including many years of nationalistic education.

The terrorist group mimics the family in size and in language, and that aids its cohesiveness in a manner that Leftist or nationalistic terrorist groups do not possess. It also makes the terrorist groups more resilient—harder to penetrate and to destroy—just as the family is, and suggests they must be targeted in unique ways—unlike "traditional" Leftist or nationalistic terrorist groups such as the Red Army Faction or Irish Republican Army.

Finally, while Islamic fundamentalist terrorist groups are more than willing to provide indoctrination to young people in an ersatz family environment, this is only part of the equation. To make terrorists, both the group and the individual are needed. Islamic fundamental terror groups capitalize on the needs of individuals, specifically, a fundamental human need for a belief system. Young people often seek indoctrination into a belief system through a religion or educational system, or the combination of the two in madrasas (Roy 2004, 92–97); they want to belong to a larger force entity for reasons described above, and for deeper, evolutionary reasons as well—as evolutionary theorist E. O. Wilson argues: "Human beings are absurdly easy to indoctrinate—they *seek* it" (1975, 562 emphasis original). Three factors cause this ease of indoctrination. First, survival in an anarchic and danger-ous world dictates membership in a group and produces a fear of ostracism from it. Second, an acceptance of or conformity to a particular status quo lowers the risk of conflict in a dominance hierarchy. Third, conformity helps keep groups together (Campbell 1972). If group conformity becomes too weak, the group could fall apart and then die out because of predation from its own or another species (Alexander 1979, 64). So, for humans, belonging to the group is better—it increases chances of sur-vival—than existing alone, even if belonging requires subordination. Thus, these factors allow terror groups to indoctrinate young people readily. They also suggest that solutions to their terror must include preventing their message from reaching young men, and replacing their message with an alternative.

## Solutions to Islamic Fundamentalist Terrorism

Understanding the causes of Islamic fundamentalist terrorism permits solu-tions to be identified. As with the causes, my objective is to offer suggestions drawn from the life sciences to augment the many social solutions that have been identified by the Bush Administration and other parties. My analysis suggests three broad solutions, and these are summarized at the end of this chapter. Some of these are gradual changes that will take time to imple-ment, but that should not discourage countries facing this threat. We should also keep in mind that significant social change is possible quickly. For example, the women's rights movement in the West is just about 40 years old, and most change has come in the last 20 years.

The first solution is to promote women's rights in the Muslim world. This is because an increase in women's rights, and particularly increases in education levels and reproductive rights, leads to smaller families. World-wide, wealthier, more-educated women have fewer children. This will lead to a smaller population and increased respect for women in the Muslim world, which will weaken the message of the terrorists over time. Of course, in the short run, major social change may create resistance, which, para-doxically, may increase the attractiveness of the Islamic fundamentalist mes-sage. This is why increasing women's rights, including for education and reproduction, must be made, although progress in these areas may be accomplished in a stepped fashion to reduce resistance to them. Second, advancing women's rights should be made in concert with other measures, such as building democracy. Third, the messenger is critically important to accomplish these goals—those advancing these profound changes must themselves be Muslims of both sexes.

Second, building democratic, liberal governments in the Middle East is critical. This is not only because it coincides with advancing women's rights and because constitutional democracy is a better form of government than authoritarianism because it ensures that the rights of individuals are protected. Equally important is that a democratic, liberal government increases the secularization of Muslim societies and promotes a competing form of government, secular and liberal, to the alternative offered by the fundamentalists.

The Bush Doctrine, the effort to foster greater political, economic, and social freedoms in the Middle East, is an important component of the solution to the terrorists. It is often reduced to the idea of spreading democracy in the Middle East. But that is not quite right. With respect to its political objectives, it would be more accurate to say that the Bush Doctrine seeks to spread effective democracy in the Middle East (White House 2006, 4–8). Effective democracy means instilling key elements of liberalism in the governments of the Middle East, Iraq most importantly. These elements include economic, social, and political freedom, including freedom of conscience, speech, assembly, and the press, the rule of law, and an independent judiciary, and respect for individual rights and human dignity. This will take place over time, President Bush and senior members of the administration often submit that these changes will occur through a generational commitment to the countries of the Middle East.

At the same time, the Bush Doctrine recognizes that there will be considerable resistance to the attempt to bring about such a transformation. Historically, this has been the case in each wave of democratization, whether it was the first that swept Great Britain and the United States, the second in Europe and Japan, the third in East Asia and Latin American, or the fourth in the Middle East. Consequently, the administration focuses its energy on elite opinions in the respective countries, knowing they will shape mass opinion. Naturally, this process will be arduous and take time, and there will be setbacks. However, with respect to Afghanistan and Iraq, effective democracy was written into their constitutions in an effort to hasten its adoption in those countries.

While the term "Bush Doctrine" will not outlast its administration, under different names its goals will guide the foreign policies of the administration's successors. Placing it in historical perspective, the Bush Doctrine is the latest incarnation of the Jeffersonian or Wilsonian impulse to advance the ideology of the United States in the world—a purpose completely shared by Democrats and Republicans alike. Recall that the Clinton administration was able to bring change to the governments of the former Yugoslavia and Indonesia.

In the Middle East, building effective democracy will cause significant economic, political, and social changes—the rise of individualism, democratic political processes, capitalism, women's rights, gay rights, the rise of secularism—that will bring Islamic states in line with the West, which will, in turn, weaken the message of the terrorists and the ability of the terrorists to attract new members and maintain their existing numbers. Effective democracy will also provide a social-leveling force, yielding more opportunity and more equitable social conditions over time that will supply other paths to high social status, and thus mates and well-being for relatives. It will provide an alternative to terrorism for some individuals.

However, the emphasis on building effective democracy in the Middle East must be made on a case-by-case basis so as not to destabilize existing allied regimes. Again, the messengers must be Muslims. Thankfully, there are many Muslim liberal reformers in the Islamic world who are allies in this endeavor—from Morocco to Indonesia and from Turkey to Pakistan.

Third, the attractiveness of the terrorists' message must be combated directly. My analysis suggests how competing messages may be tailored to accomplish this. The first of these steps is to promote the message that the family is all of Islam, not the terrorist group, and that there is no justification in the Qur'an or hadith to kill other family members, as the Islamic fundamentalist terrorists do. Additionally, public messages should emphasize that terrorists seduce individuals away from their true familial obligations, and this is un-Islamic. Shame is a powerful personal tool to combat the ideology of the terrorists. The message might be: "Usama bin Laden is a Deadbeat Dad," or "A Terrorist is a Deadbeat Son." Finally, since suicide terrorism may be thought of as sacrificing for increased mating opportunities or as a strategy to increase the social status and attractiveness of surviving relatives as mates, this lure of suicide terrorism must be combated by introducing a competing negative message about suicide bombers. This should be to suggest that the suicide terrorist failed his obligation to serve his family in this world; that suicide terrorists were filled with hate rather than the Islamic message of peace—Muslims should work for peaceful change and through irenic means to achieve their objectives; and that the suicide terrorist is mentally ill or bestial. Further, governments worldwide should put pressure on their media not to lionize suicide terrorists and to give their surviving relatives the ability to communicate. Rather, when suicide terrorism occurs, they should advance one of the competing themes.

## Final Considerations

Viewing the problem of Islamic fundamentalist terrorism through the lens of evolution and the life sciences also allows us to understand the great strengths of the West. Indeed, the victory of the West in the war against Islamic fundamentalist terrorism begins by recognizing that terrorist organizations, whether secular or religious, not only can be defeated but indeed often are. Almost all of the left-wing terrorist organizations of the Cold War were soundly defeated—from the Weather Underground in the United States to the Japanese Red Army, the Red Army Faction in Germany, and the Red Brigades in Italy. The British fought the Irish Republican Army to a standstill. The Algerians have successfully suppressed the Armed Islamic Group (GIA). The Israelis have fought Hamas and Palestinian Islamic Jihad to a stalemate. They also defeated the Palestinian Liberation Organization, as did the Jordanians. The Egyptians have broken the back of the Islamic Group and of Egyptian Islamic Jihad. Turkey has defeated the New PKK (Kurdistan Workers' Party). The Philippines destroyed Abu Sayyaf, and states throughout Southeast Asia have neutralized Jamaah Islamiyah. So while it is true that Islamic fundamentalist terrorist groups should not be underestimated—they are motivated, competent, and resilient—they have real vulnerabilities and can be defeated; indeed, they often are by governments in the Middle East and Southeast Asia.

This chapter has provided some insights into the causes and solutions to Islamic fundamentalist terrorism from the perspective of the life sciences, which is not the traditional lens to study the problem of terrorism. Such a lens yields insights into the causes and solutions to this form of terror. Finally, we should never become complacent about our foe. The essence of evolution is action, reaction, and so on. Its lesson in the present context is that a formidable, adaptive enemy requires evolving and multifaceted solutions.

ACKNOWLEDGMENTS

I would like to thank NCEAS, Raphael Sagarin, Terence Taylor, and all of the participants of the "Darwinian Homeland Security" group, especially Dominic Johnson, Kevin Lafferty, Katherine Smith, and John Tooby, for their support and insightful comments. I thank Jason Wood for his research assistance.

## Appendix: A Summary of Solutions and Research Agenda

A Summary of Solutions

1. Promote women's rights. Women's rights, and particularly increasing levels of education and reproductive rights, lead to smaller families. Worldwide, wealthier, more-educated women have fewer children. This will lead to a smaller population and increase respect for women in the Muslim world, which will weaken the messages (e.g., regarding polygamy, virility, and other behaviors) of the terrorists.

2. Support the Bush Doctrine's emphasis on building effective democracy in the Middle East, on a case-by-case basis, so as not to destabilize existing pro-American regimes. Effective democracy will bring about economic and social changes—the rise of individualism, democratic political processes, capitalism, women's rights, gay rights, secularism—that will bring Islamic states in line with the West, which will, in turn, weaken the message of the terrorists and their ability to attract new members. Effective democracy is also a social-leveling force, providing more opportunity and more equitable social conditions over time that will provide other paths to high social status and thus mates and well-being for relatives. It is thus an alternative to terrorism.

3. Terrorists mimic the family through their language; widespread use of the filial "brother" or "cousin" is often used to instill and maintain loyalty. Increasing the sense of the individual and individual rights is a mechanism to combat this, so is emphasizing in public diplomacy that terrorists seduce individuals away from their true familial obligations, and to do this is un-Islamic. Shame is a powerful personal tool to combat the message of the terrorists: "A Terrorist is a Deadbeat Dad, or Deadbeat Son." Additionally, emphasis should be placed on combating the high social status gained through terrorism, and suicide terrorism in particular, by introducing a competing, negative, message about suicide bombers and other terrorists. Elements of this message might be that the suicide terrorist failed his obligation to serve his family and that Muslims should work for peaceful change.

## Issues and Questions for Further Research

This list is intended to be suggestive and heuristic, not comprehensive.

1. Intelligence analysts should think in terms of biologically informed solutions, as well as the social solutions to defeat the terrorists. This approach provides insights not otherwise captured and would insulate them from the frequently voiced criticism of having preconceived biases or an unwillingness to incorporate new ideas and approaches that contribute to intelligence failures. In this vein, analysts should conceive of suicide terrorism as a form of inclusive fitness. The sacrifice of the individual profits genetic relatives through pecuniary gain or rise in social status for surviving relatives. This is entirely comprehensible to the intelligence analyst with an understanding of biology, who may, in turn, derive more effective counterterror policies.

2. As a contribution to intelligence assessments and prediction, examine if most male terrorists or male suicide terrorists come out of sexually skewed local populations, areas where there are fewer mates available for suicide terrorists.

3. Assessments should be made on how to decrease the social and sexual attractiveness of the suicide bomber's surviving relatives through media and film. The objective is to design competing messages that will delegitimize the behavior.

4. Assessments must be conducted concerning what textual references from the Qur'an and hadith should be employed against Islamic fundamentalist terrorists to ensure that the "family" incorporates the totality of the Muslim world, and that peaceful means should be used rather than terrorism to effect change. With respect to the former message in particular, the conception of family should be gradually broadened to incorporate non-Muslims in time.

NOTES

1. Although al Qaeda is the most famous Islamic fundamentalist terrorist organization, it is far from the only one. In fact, it is best to conceive of a family of groups, of which there are about 40 directly affiliated with or trying to imitate al Qaeda. The most dangerous of which are principally based and recruit in Europe and target within Europe, Russia, and Central Asia; these include  Al-Takfir wa al-Hijra (excommunication and exile); Groupe Salafist pur le Prediction et la Combat (GSPC, Salafist Group for Preaching and Combat); and Hizb ut-Tahrir al-Islami (Islamic Party of Liberation).

2. Evolutionary theorist and entomologist E. O. Wilson first called attention to the value of the consilient approach in 1999 (Wilson 1999). I adopted this approach to further two key issues in international relations: the understanding of war and ethnic conflict (Thayer 2004).

3. Indeed, in an otherwise exceptional compilation, there is no mention of life sciences by Gambetta (2005).

4. Here I am emulating conceptually the distinction between ultimate and proximate causes used in the life sciences. Proximate causal analysis explains many facts about an animal: why or how hormonal or stimulus-specific factors operate within it, its particular features or physiology, and the specific situations in which it demonstrates the trait. If we want to understand why birds fly south for the winter, an ultimate causal explanation of bird migration could consider factors that contribute to fitness, such as the availability of food, mates, and predators at both the indigenous and wintering areas. Ultimate or evolutionary causations are past events or processes that changed the genotype, or particular circumstances in past populations that led to the selection of the trait.

5. There is no single Salafi movement or sect, and it is more appropriate to think of it as a movement heavily informed by the thought of Sayyid Qutb. Insight analyses of Salafiyya and Wahhabism are provided by Moussalli (1999) and Algar (2002).

6. Stanley Milgram also notes the importance of what he terms "dominance structures." He argues that the "potential for obedience is prerequisite of . . . social organization" (1974, 124). Many people in his famous experiments defined themselves as "open to regulation by a person of higher status. In this condition the

individual no longer views himself as responsible for his own actions but defines himself as an instrument for carrying out the wishes of others" (1974, 134). Ethological studies have confirmed that a hierarchical dominance system within a primate band minimizes overt aggression and that it increases when the alpha male is challenged. See the excellent overview by Knauft (1991).

7. The Qur'an's ninth sura is the classic text on jihad. For an exceptional overview of the relevant sections of the Qur'an and hadith that concern jihad, see work by Bostom (2005).

## REFERENCES

Alexander, Richard D. 1979. *Darwinism and human affairs.* Seattle: University of Washington Press.

Algar, Hamid. 2002. *Wahhabism: A critical essay.* Oneonta, NY: Islamic Publications International.

Bloom, Mia. 2005. *Dying to kill: The allure of suicide terror.* New York: Columbia University Press.

Bostom, Andrew G. 2005. *The legacy of jihad: Islamic holy war and the fate of non-Muslims.* Amherst, NY: Prometheus Books.

Burke, Jason. 2003. *Al-Qaeda: Casting a shadow of terror.* London: I.B. Taurus.

Campbell, Donald T. 1972. On the genetics of altruism and the counter-hedonic components in human culture. *Journal of Social Issues* 28:21–37.

Fick, Nathaniel. 2005. *One bullet away: The making of a marine officer.* New York: Houghton Mifflin.

Frank, Robert H. 1985. *Choosing the right pond: Human behavior and the quest for status.* New York: Oxford University Press.

Gambetta, Diego, ed. 2005. *Making sense of suicide missions.* New York: Oxford University Press.

Gerges, Fawaz A. 2005. *The far enemy: Why jihad went global.* New York: Cambridge University Press.

Hafez, Mohammed M. 2003. *Why Muslims rebel: Repression and resistance in the Islamic world.* Boulder, CO: Lynne Reinner.

Harel, Amos. 2004. Would-be teen suicide bomber indicted. *Ha'aretz.* April 19. Online edition.

Kanazawa, Satoshi. 2005. Why are most suicide bombers Muslim? Unpublished manuscript. Interdisciplinary Institute of Management, London School of Economics, London.

Khosrokhavar, Farhad. 2005. *Suicide bombers: Allah's new martyrs,* trans. David Macey. London: Pluto Press.

Knauft, Bruce M. 1991. Violence and sociality in human evolution. *Current Anthropology* 32: 391–409.

Lopreato, Joseph. 1984. *Human nature and biocultural evolution.* Boston: Allen and Unwin.

McEachron, Donald L., and Darius Baer. 1982. A review of selected sociobiological principles: Application to hominid evolution. II. The effects of intergroup conflict. *Journal of Social and Biological Structures* 5: 121–139.

Milgram, Stanley. 1974. *Obedience to authority: An experimental view.* New York: Harper and Row.

Moussalli, Ahmad S. 1999. *Moderate and radical Islamic fundamentalism: The Quest for modernity, legitimacy, and the Islamic STATE.* Gainesville: University Press of Florida.

Moyer, K. E. 1987. The biological basis of dominance and aggression. In *Dominance, Aggression and War*, ed. Diane McGuinness, 1–34. New York: Paragon House.

Roy, Olivier. 2004. *Globalized Islam: The search for a new ummah.* New York: Columbia University Press.

Sprinkle, R. H. 2006. A century against the cults. Unpublished manuscript. School of Public Policy, University of Maryland, College Park.

Stern, Jessica. 2003. *Terror in the name of God: Why religious militants kill.* New York: HarperCollins.

Thayer, Bradley A. 2004. *Darwin and international relations: On the evolutionary origins of war and ethnic conflict.* Lexington: University Press of Kentucky.

Tooby, John. 2005. Remarks at the Darwinian Homeland Security workshop, NCEAS, Santa Barbara, CA, December 10.

Waltz, Kenneth N. 1979. *Theory of international politics.* Reading, MA: Addison-Wesley.

White House. 2006. *The national security strategy of the United States of America.* Washington, DC: The White House.

Wilson, David Sloan. 2002. *Darwin's cathedral: Evolution, religion, and the nature of society.* Chicago: University of Chicago Press.

Wilson, E. O. 1975. *Sociobiology: A new synthesis.* Cambridge, MA: Harvard University Press.

Wilson, Edward O. 1999. *Consilience: The unity of knowledge.* New York: Vintage.

# Chapter 9

# THE POWER OF MORAL BELIEF

SCOTT ATRAN

Suicide attacks have grown exponentially in recent years, and while they account for only 5% of terrorist events, they result in roughly 50% of the casualties due to terrorism (Atran 2006b). This poses a seemingly difficult question for an evolutionary analysis of security: how would such a self-destructive behavior not only survive, but thrive, especially in a world of limited resources? Villarreal (this volume) proposes some explanations for the evolutionary roots of self-destructive individual behaviors, suggesting that they are intricately tied to the origins of group behaviors, which manifest themselves in humans as belief systems. Here, I briefly outline the consequences of moral beliefs for terrorist and antiterrorist activities. I suggest that decision making based on moral beliefs or values creates risk/reward pathways that fall outside typically used risk-assessment frameworks based on economics, utility, or rational decision making.

Importantly, adherence to a set of moral values confers benefits to the groups to which the believing individuals identify. Value-committed people show willingness to invest in the future and delay immediate gratification (e.g., medical students who work and study intensely without initial reward). So, unlike most regular army and police, terrorist groups comprising value-committed members may be able to survive on fewer resources. Actions such as suicide bombing may further serve as "costly signals" that give the groups an appearance of vigor and robustness (see also Sosis and Alcorta, this volume). Thus, moral value–driven group associations create a positive feedback that increases group attractiveness at a low resource cost.

As a result, rather than relying on conscription or forced participation, which is often the last resort of resource-depleted and faltering organizations (e.g., various paramilitary fighting organizations in Africa, and Japanese "kamikaze" pilots at the end of World War II), jihadist organizations have been successful, and selective, in engaging new martyrs. Indeed, there is no recruitment of would-be martyrs, only a bottom-up enlistment process. As Sheikh Hamed al-Betawi, the spiritual guide of Hamas, recently summarized: "Our people don't own airplanes and tanks, only human bombs. Those who carry out martyrdom operations are not retarded, not hopeless, not poor, but are the best of our people. They do not flee from life. They are educated, not illiterate, successful in their lives" (author's interview, September 2004, Nablus, West Bank). Demographic profiling and interviews support these claims. Would-be suicide bombers are not desperate, nor are they necessarily wooed by promises of prosperity in the afterlife.

Diffusing the threat of suicide bombers will require a better understanding of the decision-making calculations that result in a person choosing the jihaddist pathway. Instrumental decision making involves strict cost-benefit calculations regarding goals and entails abandoning or adjusting goals if costs for realizing them are too high. Although the field of judgment and decision making has made enormous progress, much more is known about *economic* decision making than about *morally motivated* behavior. Indeed, a common criticism of a cost-benefit framework, which may for example put a dollar value on human health outcomes, is that it doesn't deal well with moral or ethical decisions. Further, there is little knowledge, study, or theoretical discussion of what we term "morally essential" or "sacred" values, which differ from instrumental values of realpolitik and the marketplace because they incorporate ethical (including religious) beliefs that can drive action independently of its prospect of success (Atran 2006a).

Current approaches to resolving resource conflicts or countering political violence assume that adversaries make instrumentally rational choices. However, recent work by myself and colleagues suggests that culturally distinct value frameworks constrain preferences and choices in ways not readily translatable (fungible, substitutable) across moral frameworks. Standard political and economic proposals for resolving longstanding conflicts (e.g., just material compensation for suffering) may not be the optimal when conflicts involve clashes of essential values. In particular, adversaries in violent political conflicts often conceptualize the issues under dispute as sacred values, such as when groups of people transform land from a simple resource into a "holy site" to which they may have noninstrumental moral commitments.

Nowhere is this issue more pressing than in the Israeli-Palestinian dispute, which the majority of people in almost every country surveyed (e.g., in the June 2006 Pew Global Attitudes Survey, see: http://pewglobal.org/) consistently view as the greatest danger to world peace. Our research team—including psychologists Jeremy Ginges and Douglas Medin, and political scientist Khalil Shikaki—conducted studies indicating that instrumental approaches to resolving political disputes are suboptimal when protagonists transform the issues or resources under dispute into sacred values. We found that emotional outrage and support for violent opposition to compromise over sacred values is (1) is not mitigated by offering material incentives to compromise but (2) is decreased when the adversary makes materially irrelevant compromises over his own sacred values.

In a survey of Jewish Israelis living in the West Bank and Gaza (settlers, $N = 601$) conducted in August 2005, days before Israel's withdrawal from Gaza, we randomly presented participants with one of several hypothetical peace deals. All involved Israeli withdrawal from 99% of the West Bank and Gaza in exchange for peace. We identified a subset of participants (46%) who had transformed land into an essential value; they believed that it was never permissible for the Jewish people to "give up" part of the "Land of Israel" no matter how extreme the circumstance. For these participants, all deals thus involved a "taboo" trade-off. Some deals involved an added instrumental incentive, such as money or the promise of a life free of violence ("taboo+"), while in other deals, termed "tragic," Palestinians also made a taboo trade-off over one of their own sacred values in a manner that neither added instrumental value to Israel nor detracted from the taboo nature of the deal being considered. From a rational perspective, the taboo+ deal is improved relative to the taboo deal, and thus violent opposition to the tragic deal should have been weaker. However, we observed the following order of support for violence: taboo+ > taboo > tragic; where those evaluating the tragic deal showed less support for violent opposition than the other two conditions. An analysis of intensity of emotional outrage again found that taboo+ > taboo > tragic; those evaluating the tragic deal were least likely to report anger or disgust at the prospect of the deal being signed.

These results were replicated in a survey of Palestinian refugees ($N = 535$) in Gaza and the West Bank conducted in late December 2005, one month before Hamas was elected to power. In this experiment, hypothetical peace deals (see supporting online materials) all violated the Palestinian "right of return," a key issue in the conflict. For the 80% of participants who believed this was an essential value, we once more observed that for violent opposition the order between conditions was taboo+ > taboo > tragic, where those evaluating a tragic deal showed lowest support for violent opposition. The same order was found for two measures ostensibly unrelated to the experiment: (1) the belief that Islam condones suicide attacks; and (2) reports of joy at hearing of a suicide attack (there is neuroimaging evidence for joy as a correlate of revenge). Compared to refugees who had earlier evaluated a taboo or taboo+ deal, those who had evaluated a tragic deal believed less that Islam condoned suicide attacks; and were less likely to report feeling of joy at hearing of a suicide attack. In neither the settler nor the refugee studies did participants responding to the tragic deals regard these deals as more materially likely or implementable than participants evaluating taboo or taboo+ deals. Further studies with 750 Hamas members and non-Hamas controls in June 2006 showed similar results. Follow-up interviews with Middle East leaders based on these results suggest that resolution of seemingly intractable quarrels arising from conflicting sacred values require concessions that acknowledge the opposition's core "existential" concerns (Atran, Axelrod and Davis, 2007).

These experiments reveal that in political disputes where sources of conflict are cultural, such as the Israeli-Palestinian conflict or emerging clashes between the Muslim and Judeo-Christian world, attempts to lessen violent opposition to compromise solutions can backfire by insisting on instrumentally driven trade-offs and rational choices, while noninstrumental symbolic compromises may reduce support for violence.

In promoting alternative pathways for those inclined toward suicide operations, the key is not to try to undermine sacred values that inspire people to radical action, or attempt to substitute one's own preferred values by forceful imposition or through propaganda. Studies in cognitive and social psychology show that such tactics usually only incite further moral outrage and extreme behavior. Rather, the aim should be to show how deeply held sacred values can be channeled into less belligerent paths. In interviews with mujahedin who have rejected suicide bombing, I find they remain very committed to Salafi principles and their religion remains rock steady and deep. But those who seem to best succeed in convincing their brethren to forsake wanton killing of civilians do so by promoting alternative interpretations of Islamic principles that need not directly oppose Salafi teachings.

Preventative efforts should engage cultural orphans in cities and in cyberspace with alternate peer groups that are as committed and compassionate toward their own members as are terror groups. This requires careful monitoring, rather than simply taking down existing jihadi Web sites. What is needed is subtle infiltration of opportunities to create chat rooms and also physical support groups that advance causes that can play to jihadi sentiments but that are not destructive, for example, providing faith-based social services. Another possibility is to set up parallel networks (even sports clubs) that can siphon off action-oriented young adults before a downward spiral sets in, where core meaning in life is sought, and found, in religious networks that sanctify vengeance at any cost against stronger powers, even if it kills the avenger.

## REFERENCES

Atran, S. 2006a. Devoted actor versus rational actor models for understanding world conflict. Presented to the National Security Council at the White House, September 14, 2006. Available at http://cstsp.aaas.org/content.html?contentid=621.

Atran, S. 2006b. The moral logic and growth of suicide terrorism. *Washington Quarterly* 29 (Spring): 127–147.

Atran, S., R. Axelrod, and R. Davis. 2007. Sacred barriers to conflict resolution. *Science* 317:1039–1040.

Ginges, J., S. Atran, D. Medin, and K. Shikaki. 2007. Sacred bounds on the rational resolution of violent political conflict. *Proceedings of the National Academy of Sciences USA* 104: 7357–7360.

Part Five

# ECOLOGY AND SECURITY

Chapter 10

# FOURTEEN SECURITY LESSONS FROM ANTIPREDATOR BEHAVIOR

DANIEL T. BLUMSTEIN

Knowledge is power, whether it creates new ways to control a situation, or whether it simply explains the biological basis of a situation. I believe that there are lessons about security that we can learn from the sheer diversity of ways that nonhumans avoid predation. I am a behavioral ecologist. Behavioral ecologists adopt an economic approach when we study animals in natural settings to understand the evolution, diversity, and maintenance of behavior. We expect that costly behaviors will be selected against, unless there are overwhelming benefits associated with them. We expect animals will make fundamental trade-offs in how they allocate time and energy, and that over evolutionary time those that make the proper trade-offs will persist, while others will go extinct. Below I derive 14 lessons from the study of antipredator behavior that are relevant to designing security systems to manage terrorist threats, dealing with insurgencies, as well as managing ongoing biosecurity challenges.

We can make sense of the diversity of antipredator behavior several ways. One useful way focuses on the predatory sequence (Endler 1986; Caro 2005). Predators encounter potential prey and must identify them as suitable. Then, they must approach and attack the prey, prevent them from escaping, and consume them. Antipredator defenses may work at any of these steps. With this predatory sequence in mind, we can examine the interactions between predators and prey. For instance, prey should engage in behaviors that reduce detection by predators: they should be cryptic, or active at times when predators are not around. Prey may engage in group defenses. A commonly hypothesized benefit of sociality is to reduce the risk of predation by either increasing the ability of prey to detect predators, or simply spreading the risk among more individuals (Krause and Ruxton 2002). Once detected, prey should make themselves less profitable to predators. Increasing the cost of attacking or handling prey is an effective means to decrease profitability (Krebs and Davies 1993). For instance, many species have spines or other defensive structures. Remember, from the predator's perspective, it is looking for a meal, so the economics of handling a difficult prey will reduce the prey's profitability. Prey may also communicate to predators that they have been detected (Blumstein 2007). Such detection signaling works when predators require stealth for success. When a prey signals that it has identified a predator, the predator's hunting success has just been massively decreased. Prey may gang up and attack a predator. Such mobbing is common in a variety of species, and it often successfully moves the predator away from a particular location. Individually fighting a predator, or fleeing from a predator is an action of last resort. Flight is both energetically costly and involves an opportunity cost in that animals must stop what they have already been engaged in. Fighting is risky and therefore something that should be avoided at all costs.

A common behavioral ecological paradigm envisions individuals trading off the probability of starvation versus the probability of predation (Mangel and Clark 1988). Imagine a refuging rodent, for example, a kangaroo rat that lives in the desert, or a marmot in an alpine meadow. It lives in a burrow but must come out to forage. If it remains in its safe burrow, it will ultimately starve to death. If it emerges, it faces some risk of predation.

Our first lesson is that *avoiding all risk is impossible.* Virtually all animals must live with some risk of predation at some point in their lives. What cues should they use to assess risk? How should they modify their foraging behavior based on these cues?

Animals may use direct or indirect cues of risk. Direct cues are those by which individuals detect a predator. Indirect cues are those where there is some probabilistic relationship with predation. For instance, if predators hide in dense cover, avoiding dense cover might be an effective way to avoid predation. Regardless of whether they are direct or indirect, all cues vary in their reliability. For instance, if a predator is sighted, the prey knows that it is around, whereas if the scent of a predator is detected, the predator may be around now, or it may have been around at some previous time. Thus, prey often face the problem of estimating the reliability of a particular cue or situation.

Our second lesson is that *overestimating risk is a good strategy in many circumstances.* Theoretical models (Bouskila and Blumstein 1992) suggest that when faced with a starvation-predation risk trade-off, and imperfect information about the true risk of predation, being conservative—that is, overestimating risk—is an optimal strategy. Thus, we expect that animals may use rules of thumb to estimate predation risk and that these rules should be systematically biased toward overestimating the risk of predation. Clearly, assuming there is a 100% probability of being killed will be nonadaptive (an individual will then starve to death), but a modest degree of overestimated risk can be adaptive.

Given that individuals must take risks, how can risks be avoided? Imagine you find yourself at an automatic teller machine (ATM) in the middle of a bad neighborhood. One strategy to reduce risk would be to approach the ATM cautiously and spend a lot of time looking around while there. By doing so, you will spend more time in an exposed position. An alternative strategy would be to run in and run out as quickly as possible. We see evidence that animals use both strategies in nature. Some species are more vigilant in risky areas, while others are less vigilant, and by being less vigilant, they are able to reduce their exposure to predators because they decrease the amount of time in risky areas. Our third lesson is thus that *it is possible to reduce risk by limiting exposure or by being very careful when in risky areas.*

When prey detect a predator, many species produce obvious visual, acoustic, or olfactory signals referred to as alarm signals. Such alarm signals may be directed to the predator (to signal detection) or to other members of one's species (to warn them about the presence of the predator). There is considerable interest in understanding the evolution and adaptive utility of such signals because they are obvious and potentially risky (they may expose individuals to some risk of predation if the predator decides to focus on the signaler). That said, we see that alarm signals are often produced only when individuals are safe. The marmots I study typically run back to a burrow before alarm calling. The wallabies I have studied foot thump only when I am not walking directly toward them (and thus, when they perceive themselves as not being directly targeted). Thompson's gazelles, an African ungulate, stott (bounce up and down in front of predators with a characteristic stiff-legged gait), but only once they are a certain distance away from risky predators (Caro 2005). Historically, in rodents at least, alarm calls evolved to signal detection to the predator (Shelley and Blumstein 2005). Such calls have subsequently been "exapted" to have a conspecific-warning function. An exaptation is a trait that initially evolved for one function but later has been adapted to a new function (Gould and Vrba 1982). We expect many traits to be exaptations because evolution by natural selection works from the set of existing traits. Evolution, as Francios Jakob once wrote, works by "tinkering" with what you have, rather than creating some ideal trait de novo (Jakob 1977).

Alarm signals provide our next three lessons. Fourth, *detection signaling is a good idea and may, in some circumstances, reduce predation risk by encouraging the predator to select another target.* Fifth, *individuals should signal when they are in a relatively safe position:* flee to safety then signal. Sixth, *exaptations are common.* Defenses are adapted from preexisting behaviors rather than building de novo structures. Clearly, humans can escape from this constraint in that natural selection acts on existing variation, but humans could create their own variation. In theory, we could build a completely new type of transportation vehicle, but such novel construction would itself have costs. Adapting our defenses to meet new challenges from what we already have might be an effective long-term defensive strategy. At times, however, there are novel offenses. A defining characteristic of exotic plants that successfully invaded North America is that they had novel phyotochemicals and thus were defended against resident herbivores (Cappuccino and Arnason 2006). Novel offensives may require novel defenses, but these will most rapidly evolve from preexisting defenses.

Risk assessment is imperfect, and it behooves those detecting alarm signals to pay particular attention to who is producing the signal. Imagine two individuals: Nervous Nelly, and Cool-Hand Lucy. Nervous Nelly produces alarm calls whenever she detects any movement. Nelly calls to falling leaves, rustling bushes, and sometimes, to real predators. Cool-Hand Lucy, on the other hand, is much more discriminating. She calls only when she detects a predator. Given systematic differences in the reliability of callers, it should behoove those responding to them to try to assess caller reliability. In many species, information from unreliable individuals is discounted. Thus, Nervous Nelly (who often "cries wolf") is ignored. In some species, however, reliability assessment works in a slightly different way. Reliable individuals that make mistakes are given the benefit of the doubt, and unreliable individuals, probably because they are unreliable, elicit considerable independent investigation (Blumstein et al. 2004b).

Our seventh lesson is that *it is very important to assess signaler reliability* and to behave accordingly. However, how one behaves may vary in that unreliable individuals may be discounted or elicit more independent investigation. Our eighth lesson is that *there can be qualitatively different responses to the same situation or problem* and that these different strategies may all ultimately be successful strategies. Evolution provides us with a plethora of ways to respond to risk.

When faced with an uncertain risk, what should individuals do? Many species inspect their predators to better estimate the true risk of being around them (Caro 2005). Clearly, predators are not always hunting. Satiated predators may have obviously full bellies (envision a lion resting after a large meal). Acquiring more information about the true risk of predation is important because it is relatively costly to flee a nonhunting predator.

Our ninth lesson, thus, is that *reducing uncertainty by predator inspection is an important way to reduce costly responses.* Individuals that engage in costly responses indiscriminately will be selected against, while those that respond in an optimal way by responding appropriately to risk may do better in the long run.

A common theme throughout this chapter is avoiding costly responses. And, nonhuman antipredator behavior provides several insights into how this can be achieved. In many cases, costly antipredator behaviors are lost when there are no longer any predators around. For instance, many species become less cryptic, less wary, or otherwise lose antipredator behavior when isolated on islands or other predator-free locations (Blumstein and Daniel 2005). Such responses are adaptive in that by no longer allocating energy or time to antipredator behavior, animals are able to allocate energy or time to other activities.

Our tenth lesson is that it is adaptive to *reduce defenses when risk decreases.* Maintaining high vigilance when there is truly no risk is costly, and individuals doing so will be out-competed by those that reduce defenses adaptively.

However, there are some situations where we see evidence of antipredator behavior persisting despite long periods of relaxed selection (i.e., isolation from predators). One hypothesis explaining this is something I have called the multipredator hypothesis (Blumstein 2006). Most species have more than a single predator, and this realization has several important implications. First, we might expect generalized defenses that work well against more than a single predator. Second, being able to survive one predator but not another does not make much sense, and we might, therefore, expect "packages" of antipredator behavior. Importantly, we should not expect the various traits that constitute an effective defense to assort independently. For instance, consider a baby ungulate, such as a pronghorn antelope, which relies on crypsis and immobility to survive. A cryptic pronghorn who bounces around wildly, would be killed by coyotes or eagles. Thus, these traits should not be independent. Similarly, an animal that is exposed to both eagle predation and wolf predation should not have their antipredator traits be entirely independent (surviving one only to be killed by the other would not be favored by evolution). At the genetic level, we may see evidence of genetic correlations, and we might expect linkage (whereby traits are colocated on chromosomes to resist independent assortment). Studies of kangaroos and wallabies provide some support for the multipredator hypothesis: species living without any predators quickly loose all antipredator behavior, but those living with at least one predator maintain antipredator abilities for other, absent, predators (Blumstein et al. 2004a).

Several lessons come from thinking about the multipredator hypothesis. Our eleventh lesson is that we should aim to *have generalizable defenses that work against more than a single threat.* Our twelfth lesson is that *unless there are great costs to maintaining a defense in the absence of a specific predator, it may be a good idea to maintain all defenses.* This of course, focuses us on estimating the costs of maintaining apparently no longer useful defenses. In some cases we should expect defenses to be independent, but in other cases, we should expect them to be linked.

There are several ways by which prey could not respond to ongoing nonthreatening situations. The multipredator hypothesis focuses on the evolutionary loss of antipredator behavior, but species can respond in much more dynamic ways based on experience. A fundamental response is to habituate to ongoing stimulation. Habituation is seen when the magnitude of responses declines with repeated nonthreatening exposures. For instance, the Arabs used habituation effectively against the Israelis in the 1973 war (Rabinovich 2004). Before attacking Israel, Egypt had 40 military exercises on Israel's borders. This led Israeli security analysis to discount the threat of a troop buildup on their border. However, habituation is not ubiquitous, and individuals may sensitize, or have higher-magnitude responses, to repeated exposures.

Maintaining vigilance in situations where there are many false alarms is difficult (see appendix). Our thirteenth lesson is that *we should often expect habituation when there are many false alarms*. Importantly, understanding the conditions under which habituation or sensitization occurs is a fundamentally important question in security studies.

Our final lesson is that *maintaining flexible responses is often a good idea*. There are clearly time, energy, and opportunity costs of assessing risk, and by having inflexible constituent defenses, individuals need not pay the costs of trying to adaptively deploy defenses. In some situations, either when the costs are low, or when there are limited benefits from proper assessment, we should expect fixed responses. However, flexible and adaptable responses that are deployed only when necessary, in the long run, may allow individuals to allocate less energy to defense and more to other important activities.

## Policy Implications

Studies of antipredator behavior illustrate remarkable flexibility of mechanisms: there is no single way to solve a particular problem. The optimal solution will likely depend on specific constraints, as well as the costs of making mistakes.

Whenever possible, we should seek to adapt defenses from preexisting resources. For instance, rather than creating a novel and unique "Department of Bioattack Detection," health care systems can be better developed and communication among hospitals and government agencies improved, so that biological attacks could be quickly detected. Importantly, an improved public health system will have positive benefits for citizens even when there are no terrorist attacks, and a strong public health system will help us respond to natural pandemics. Similarly, first responders should be given radiological and chemical weapons monitors (as some are), rather than creating an entirely new agency tasked to detect a low-probability (but admittedly high-consequence) threat.

Raising the alarm with press releases describing "credible information of threats," changes in the DEFCON level, or the Department of Homeland

Security threat level should be expected to communicate to both opponents and allies. Detection signaling has a long and functional evolutionary history, and this realization could be used strategically. However, to use it strategically, we must better understand the conditions under which we habituate to repeated false alarms. Without a fundamental understanding of this, we are likely to have the unintended consequence of habituating those that we need to remain vigilant. Importantly, when specifically targeting opponents, it is important to signal from a position of safety. In a policy context, this would mean we signal when we are certain that there is something to signal about. Thus, it is essential to choose when to signal to opponents; crying wolf has its own costs!

Faced with uncertain threats, information acquisition is essential, and this is a fundamental characteristic of adaptive, flexible responses. This, itself, is not a novel suggestion, but a behavioral ecological perspective does emphasize the costs and benefits of information acquisition and response are key things to evaluate when selecting a response. If making a mistake is not that costly, we should tolerate mistakes. However, whenever mistakes are costly, we should be very discriminating. Interestingly, there seem to be two ways to respond to uncertainty: discount it or use it as an opportunity to acquire more information. For instance, the information from the FBI agent that warned about certain foreigners enrolling in flight schools contained little that was actionable. Using this as an opportunity to acquire more information, in this case, might have prevented the 9/11 attacks.

Maintaining not currently functional defenses may be, in the long run, a good idea, but we should be very sensitive to the opportunity costs of doing so. For instance, if it can be done economically, ongoing monitoring for controlled pathogens (smallpox, polio, etc.) will allow early detection and thus control. Maintaining general defenses, such as health care systems, will allow us to respond more efficiently to a biological attack.

While not novel, examining the diversity of antipredator adaptations highlights, from a perspective directly concerned with life and death, the fact that we must be comfortable living with risk. It is impossible to completely avoid risk, and by trying to do so, we will suffer other costs. One (admittedly not politically tenable) response to a major terrorist attack, like 9/11 or the attacks in Madrid and London, would be to ignore them. In all three countries, more citizens are killed by drunk drivers each year than die at the hands of terrorists. Targeting drunk driving, smoking, or obesity, rather than terrorist attacks, might be a better way to save citizen's lives. But humans overestimate the risk of large, rare events (Cohl 1997), and politicians must respond to this fear (we also are more fearful of novel events, human-made risks, risks we have no control over, uncertain risks, and risks that may kill us in shocking ways [Ropeik and Gray 2002]).

The ancillary costs of responding to terrorist threats may, in some cases, be very high. The United States struggles with the trade-off between losing individual liberties and employing sophisticated surveillance against people who might be U.S. citizens. With a fixed budget, funds allocated to defense cannot simultaneously be allocated to health care, science, or education. Thus, overreacting has its intrinsic costs.

We see clear evidence that humans recognize the multiple ways to reduce risk. Many vehicles traveling from the Green Zone to the airport in Baghdad do so at very high speeds—a time-minimizing strategy that reduces exposure. Military personnel patrolling Iraqi roads do so very carefully, examining piles of trash, dirt, and household items to determine if they hold improvised explosive devices. With experience, search images develop and personnel are better able to detect these hidden bombs (National Public Radio 2005). We also see that the constraints created by having to maintain a physical presence selects in the military personnel this highly vigilant behavior.

Also, while not novel, studying antipredator behavior illustrates the value and efficacy of increasing the cost of attacking (i.e., building defenses) as a viable deterrent. Of course, as airports, embassies, and government office buildings are hardened, terrorists will choose softer targets, such as marketplaces and hotels. Nevertheless, the lesson is that increasing the cost of attack is a viable defensive strategy to defend important resources.

## Future Research Needs

While nonhuman antipredator behavior creates a toolbox of strategies for those who wish to apply it to human security issues, future research should develop the following:

Models to explain the condition under which unreliable individuals will be ignored or will elicit independent investigation. Results will help us develop better strategies for responding to uncertain events.

Models to explain the conditions under which habituation or sensitization occurs. Results will help us defensively by preventing a decline in vigilance/responsiveness, and offensively to better understand how to reduce the vigilance of opponents.

Models to explain the conditions under which maintaining potentially costly and no longer useful antipredator behavior is a good strategy. Results will help us allocate scarce resources efficiently.

A better understanding of the consequences of using a general defense as opposed to a specific defense. Assuming specific defenses cost more to develop, what is the added value obtained by developing them. Results will help us allocate scarce resources efficiently.

A better understanding about how, specifically, species respond to novel
   threats. Do novel threats require truly novel responses, or can generalized or exapted responses work well?

## Appendix: An Analysis of Citizen Responses to Changes in Department of Homeland Security Threat Levels

*Daniel T. Blumstein and Elizabeth M. P. Madin*

The Department of Homeland Security (DHS) threat level system was instituted in September 2002. It has five levels: 1 = low, 2 = guarded, 3 = elevated, 4 = high, and 5 = severe. Since its institution, it has remained at
either level 3 or level 4. While it was originally developed to help responders plan for terrorist attacks, it was made public, allowing us to analyze
how the public responds to changes in threat levels. The DHS states on its
threat level Web page (http://www.dhs.gov/xinfoshare/programs/Copy_of_
press_release_0046.shtm) that "raising the threat condition has economic,
physical, and psychological effects on the nation"; however empirical evidence as to the nature of such effects remains scant. In particular, citizens
are encouraged as part of this system to utilize the Web site www.Ready.gov
and the DHS information line (1-800-BE-READY) as their primary sources
of information on preparedness for possible terrorist attacks. In order to
test the effects of this system on citizen responses, we conducted two analyses. First, using bivariate correlations, we looked at how public responses
changed over time and as a function of the specific DHS threat level. Second, we fitted a series of general linear models to isolate the effect of DHS
threat level, after explaining variation accounted for by date.

Our response measures included CNN/USA Today/Gallup polling data
(from www.pollingreport.com/terror.htm) from the question: "How worried are you that you or someone in your family will become a victim of terrorism: very worried, somewhat worried, not too worried, or not worried at
all?" This poll surveyed approximately 1000 Americans periodically since
9/11. We also included data provided to us from the DHS that summarizes
the number of hits to the DHS Web site (www.Ready.gov) and calls to the
DHS information line (1-800-BE-READY).

We found some evidence of habituation, a decline in responsiveness over
time, and no strong evidence that American citizens responded to variation
in DHS threat levels by increasing their fear (Fig. 10.1). Specifically, the per-
centage people reporting that they were "very worried" decreased as a function of the highest level DHS threat that month, and the number of people
reporting that they were "not too worried" increased. There were also strong
effects of time on responses: the number of page views to www.Ready.gov
and phone calls 1-800-BE-READY decreased over time and were not sensitive

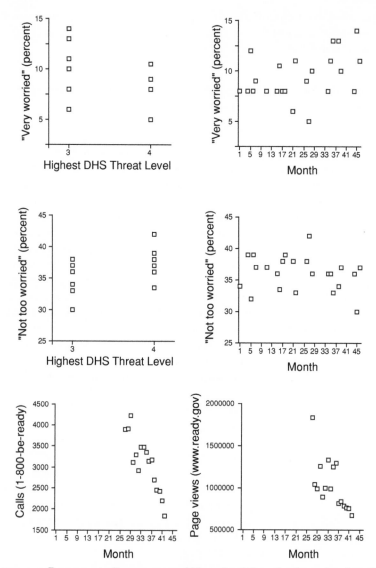

FIGURE 10.1 Response to Department of Homeland Security threat levels and over time. Month 1 = November 2001. Data for www.Ready.gov and 1-800-BE-READY prior to January 2004 were unavailable.

to variation in the highest level DHS threat that month. These results were supported by the results from the general linear models: after accounting for variation explained by month, threat level had no significant effect on page views or phone calls. Interestingly, there were no significant effects of either date or highest threat level on the responses to poll data.

American citizens seemingly obtained response to terrorist threat information early on from the DHS and stopped acquiring information over time. Changes in DHS threat level had no effect on acquiring information by going to these recommended DHS sources. We also found some evidence that the American public has habituated to DHS threat warnings since elevations in the DHS threat level led to, if anything, a decline in the number of Americans that reported themselves to be "very worried" about a possible terrorist attack. Our results highlight the point that habituation is a factor with which governments should be very concerned and to which effective homeland security systems will need to pay attention.

## ACKNOWLEDGMENTS

I thank Rafe Sagarin for inviting me to join this working group; Dominic Johnson, Rafe, and Gerat Vermeij for comments on a previous draft; and Dominic and Rafe for ongoing stimulating conversations about these and many other ideas about the links between academic behavioral ecology and security.

## REFERENCES

Blumstein, D. T. 2006. The multi-predator hypothesis and the evolutionary persistence of antipredator behavior. *Ethology* 112: 209–217.

Blumstein, D. T. 2007. The evolution of alarm communication in rodents: Structure, function, and the puzzle of apparently altruistic calling. In *Rodent Societies*, eds. J. O. Wolff and P. W. Sherman, 217–237. Chicago: University of Chicago Press.

Blumstein, D. T., and J. C. Daniel. 2005. The loss of anti-predator behaviour following isolation on islands. *Proceedings of the Royal Society London B* 272: 1663–1668.

Blumstein, D. T., J. C. Daniel, and B. P. Springett. 2004a. A test of the multi-predator hypothesis: Rapid loss of antipredator behavior after 130 years of isolation. *Ethology* 110: 919–934.

Blumstein, D. T., L. Verenyre, and J. Daniel. 2004b. Reliability and the adaptive utility of discrimination among alarm callers. *Proceedings of the Royal Society London B* 271: 1851–1857.

Bouskila, A., D. T. Blumstein. 1992. Rules of thumb for predation hazard assessment: Predictions from a dynamic model. *American Naturalist* 13:, 161–176.

Cappuccino, N., and J. T. Arnason. 2006. Novel chemistry of invasive exotic plants. *Biology* Letters 2: 189–193, doi: 10.1098/rsbl.2005.0433.

Caro, T. 2005. *Antipredator defenses in birds and mammals.* Chicago: University of Chicago Press.

Cohl, H. A. 1997. *Are we scaring ourselves to death? How pessimism, paranoia, and a misguided media are leading us toward disaster.* New York: St. Martin's Press.

Endler, J. A. 1986. Defense against predators. In *Predator-Prey Relationships,* eds. Feder, M. E. and G. V. Lauder, 109–134. Chicago: University of Chicago Press.

Gould, S. J., and E. S. Vrba. 1982. Exaptation: A missing term in the science of form. *Paleobiology* 8: 4–15.

Jakob, F. 1977. Evolution and tinkering. *Science* 196: 1161–1166.

Krause, J., and G. D. Ruxton. 2002. *Living in groups.* Oxford: Oxford University Press.

Krebs, J. R., and N. B. Davies. 1993. *An introduction to behavioural ecology,* 3rd ed. Oxford: Blackwell Scientific Publications.

Mangel, M., and C. W. Clark. 1988. *Dynamic modeling in behavioral ecology.* Princeton, NJ: Princeton University Press.

National Public Radio. Morning Edition, May 20. 2005. One of the most common dangers to American soldiers in Iraq is the improvised explosive device, or IED. Army Brig. Gen. Joseph Votel, head of a Pentagon task force working to reduce the number of IEDs in Iraq, discusses the threat. http://www.npr.org/templates/story.php?storyId=4659766.

Rabinovich, A. 2004. *The Yom Kippur war: The epic encounter that transformed the middle east.* New York: Schocken Books.

Ropeik, D., and G. Gray. 2002. *Risk: A practical guide for deciding what's really safe and what's really dangerous in the world around you.* Boston: Houghton Mifflin.

Shelley, E. L., and D. T. Blumstein. 2005. The evolution of vocal alarm communication in rodents. *Behavioral Ecology* 16: 169–177.

Chapter 11

# POPULATION MODELS AND COUNTERINSURGENCY STRATEGIES

DOMINIC D. P. JOHNSON AND JOSHUA S. MADIN

*All guerilla units start from nothing and grow.*
MAO TSE-TUNG

Efforts to contain terrorist or insurgent populations share many characteristics that would be familiar to any modern ecologist studying the dynamics
of natural populations. The quantitative tools of ecology may therefore be
useful in understanding these types of conflicts. We use data on two insurgencies, one that was defeated (Malaya 1948–1960) and one ongoing (Iraq
2003– ), to examine whether a population model can offer useful insights
about insurgency growth, and what counterinsurgency strategies are most
likely to be effective. Population models focus on parameters critical to the
success or failure of counterinsurgency campaigns: insurgent population
size, mortality rates, recruitment rates, and population carrying capacity
caused by resource constraints. After fitting a simple population growth
model with external mortality to the two data sets, we demonstrate the relative impact of seeking changes in mortality, recruitment, or carrying
capacity on the future insurgent population size, and the importance of
extinguishing threats early on, before they become established.

Our models are, like all models, simplifications of complex events, focusing on a handful of parameters that we deem important and ignoring many
others. For this reason, the results must be interpreted with caution. However, this is also precisely why models are useful: (1) the aim is to distill key
factors from a confusing array of information; (2) some of our results succeed in describing actual events in the historical Malayan insurgency, suggesting that the models can account for real patterns with only basic
variables; and (3) in fast-moving conflicts there is often little time or opportunity to obtain or verify comprehensive data, so the most useful analytical
tools are ones that offer straightforward predictions with the limited data
available. If we can extract fundamental patterns from the fog of war, we
can identify practical goals for action.

Model results for Iraq suggest that: (1) if conditions had stayed the same as they were in January 2006 (i.e., low levels of sectarian violence), the Iraqi insurgency would have collapsed in 4.5 years, but only if the United States continued its trend of improving military performance; and (2) moderate changes to recruitment, carrying capacity, or mortality could, in combination, have defeated the insurgency in 6 to 12 months. The rise of sectarian violence in Iraq may therefore have foreclosed an opportunity to defeat the insurgency in a relatively short time.

## Natural Systems as Ecological Models of Conflict

Natural systems are a hugely intricate web of interacting species, processes, environmental factors, and human-induced change. In order to distill these highly complex systems into manageable, understandable units, biologists have learned that populations are often best understood by focusing simply on fundamental data and models relevant to population change. The same may be true of apparently confusing human conflicts, but the rationale for doing so may occur only to biologists. Like populations of any species, human groups unified by an agenda or ideology are born, grow, and eventually die. Although the actual factors that regulate such human groups and nonhuman biological populations may be different, their underlying effects on population size are the same. For example, the growth of human groups is driven by immigration and recruitment; similarly, the growth of animal and plant populations is driven by immigration and biological reproduction (in ecological parlance, "recruitment" of surviving offspring into the breeding population).

Quantitative models of human conflict have a long history, beginning with Lewis Fry Richardson's and Quincy Wright's studies of arms races and the causes of war (Richardson 1960, 1978; Wright 1983). Since then, rapidly increasing computational capabilities and techniques have allowed the development of a broad range of models that may be used to study human social dynamics (Cederman 1997; Epstein 1997; Ehrlich and Levin 2005; Epstein 2006). Models derived specifically from ecology have, for example, been used in studying the dynamics of public protest and coercion (Francisco 1996, 2004), and the life cycles of empires (Turchin 2005).

Ecological models could, in principle, be generalized to any type of conflict from gang warfare to conventional interstate wars. Here, however, we focus on insurgency, exploiting the techniques and tools of a simple population model to explore the dynamics of insurgent populations. Models offer two main opportunities: (1) by understanding and manipulating population parameters such as mortality, recruitment, and carrying capacity, populations can be either sustained (as with agriculture, fisheries, etc.) or eradicated (as with invasive species, pests, etc.); and (2) population models can also offer insight into how populations can be controlled or prevented from emerging in the first place. We do not of course expect all aspects of these models to map neatly onto the complex human conflicts they represent. Instead, we aim to conduct an exploratory probe into tools that could unveil novel patterns and predictions, even with limited data.

### Characteristics of Insurgencies

Insurgencies are a phenomenon common to all cultures, all regions of the globe, and all periods of history (Galula 1964; Beckett 2001; Fearon and Laitin 2003). The latter part of the twentieth century, fresh in many people's living memory, was unusual in that insurgencies throughout the world tended to come under the banner of proxy U.S.-Soviet superpower conflict; but insurgencies have been common before, during, and after the Cold War. While the causes of insurgency vary widely, ranging through social or political reform, nationalism, ideology, religion, and power struggles, each is characterized by an initially weaker group taking up arms to challenge the ruling power. Sometimes the insurgents' goals represent those of a minority, sometimes those of a majority, and in either case the active participants may be few or many. Insurgencies also range vastly in their duration and consequences. Some fizzle out before they get going, others, such as in the conflict in Ethiopia, raged for 30 years and claimed 250,000 lives before the insurgents succeeded in establishing their own country of Eritrea in the 1990s.

Like successions of organisms on a landscape, insurgencies are more likely to emanate after a period of change or disturbance. For example, Somalia collapsed into multilateral tribal conflict after the fall of dictator Siad Barre in 1991 (Clarke and Herbst 1997). Insurgents tend to exploit windows of opportunity that arise from a power vacuum, spreading quickly or coming out into the open after having kept a low profile (as with Shi'a uprisings in Iraq after the 1991 Gulf War). Below, we examine each of the key parameters that determine population growth as they apply to insurgencies: population size, recruitment rate, carrying capacity, and mortality.

### Population Size

For an insurgent population to persist and prosper, there must be both suitable conditions, and an adequate initial population from which to grow. If a seed of dissidence is planted, but the political and social conditions are not amenable for growth, then the population will either remain small or become extinct. Communism in Greece, for example, never made any headway after World War II because it lacked salience for Greek culture and institutions, and the communists were allied with many of Greece's traditional enemies (Galula 1964). If, on the other hand, the conditions for growth are in place, but the initial population is too small, then individuals may not interact frequently enough to organize, execute their plans, or recruit new individuals, and the population may again fail to grow (cf. the "Allee effect," in which the per capita growth rate falls off at low population densities [Stephens et al. 1999]). The left-wing Weather Underground Organization, which plotted to overthrow the United States government in the 1970s, for example, started off small and progressively lost members through extremism and accidents until it disappeared. But when the conditions are right—typically during a period of political turmoil—and a large enough initial population exists that can recruit new members, an insurgency may develop.

The task of controlling or neutralizing an insurgency may at first glance seem much easier during the initial stages of population growth, when the group is vulnerable because of its small size, inexperience, low publicity, few interactions, and dispersed locations. For similar reasons, dwindling populations of endangered species experience a progressively greater risk of extinction, even if there is in theory plenty of genetic variation for a viable population (Macdonald and Johnson 2001). Counterinsurgency strategies should therefore strike fast, hard, and, importantly, early. As David Galula put it in his classic study: "The longer the insurgent movement lasts, the better will be its chances to survive its infantile diseases and to take root" (Galula 1964, 68). Unfortunately, there are three major impediments to nipping insurgencies in the bud: (1) it is precisely during this initial stage that the growing population is hardest to detect—it may lie low, fail to attract publicity, or simply fall below the radar until it has gained considerable strength; (2) even if they are noticed, it is impossible to know which of many such groups will develop into a significant threat and which will fizzle out on their own; (3) even if a genuine threat is detected early on, it may be impossible to garner the necessary political support for what remains an unfamiliar and hypothetical threat to the public: "The potential danger is enormous, but how to prove it on the basis of available, objective facts? How to justify the efforts and sacrifices needed to smother the incipient insurgency?" (Galula 1964, 64).

## Recruitment

How does population size increase? In human groups formed to carry out an insurgency, biological reproduction is negligible because it operates over a time scale of generations, whereas the time scale of insurgencies is usually measured in years (Fearon and Laitin 2003). There are exceptions in which demographics can ultimately influence the outcome of ethnic conflicts (Toft 2002). For instance, Celtic tribes fought Roman forces in Britain over several centuries, prompting permanent border garrisons and the construction of Hadrian's Wall in A.D. 122. Demographic changes in such long-term conflicts do have a bearing on the material forces that each side can commit to battle. However, the lifespan of most modern insurgencies is much shorter. As such, they must grow by recruiting existing individuals, either actively (by coercing or encouraging receptive individuals and communities) or passively (via advertising or demonstrating a salient ideology or cause). People join insurgencies for all sorts of reasons—ideology, political aspirations, revenge, to oust occupying powers, to gain status or reputation, religious calling, and so on. There are sometimes economic incentives, particularly in Islamic terrorist organizations, whose madrassas offer financial rewards and "education" for new jihadis. Sometimes economic benefits accrue to insurgents' families, such that even suicide terrorism offers some payoff to kin. Hamas has an excellent reputation among Palestinian communities for providing a wide range of social support functions. Often social incentives are also strong, encouraging active participation in insurgencies through the reverence of martyrs and the elevation of their kin's standing in the community (Stern 2003; Pape 2005; Thayer, this volume).

The key point here is that because of the numerous rationales and routes to join an insurgency, a process that is often invisible, recruitment rate is very hard to verify. As Donald Rumsfeld (2003) lamented on October 16, 2003: "Today, we lack metrics to know if we are winning or losing the global war on terror. Are we capturing, killing, or deterring and dissuading more terrorists every day than the madrassas and the radical clerics are recruiting, training, and deploying against us?" If the secretary of defense does not know, with all the resources of the U.S. intelligence services, who does? Population models provide an intriguing alternative approach to tracking variables that are hard to estimate or collect, such as recruitment rates, by building a model from more tractable and available variables, such as mortality.

## Carrying Capacity

Populations rarely grow indefinitely. Eventually some essential resource or other, such as food or space, becomes scarce, and population growth slows to a halt (or even declines, if the essential resource is depleted). The dependence of a population's dynamics on the density of its constituents is thus a central concept in ecology (Begon et al. 2005). A population grows when input (birth or immigration from other populations) is greater than output (death or emigration). As population density increases, the scarcity of a resource decreases input rate and/or increases output rate, such that the population size plateaus off at what is called the "carrying capacity." Any analogous group of resource-dependent organisms experiences the same phenomenon. Insurgencies will be subject to a similar process of density dependence due to a number of limiting factors: the maximum pool of potential insurgents, available safe areas, support and supplies from the indigenous community, logistical resources, weapons, and so on.

In Malaya, the British literally moved nearly half a million ethnic Chinese squatters out of the jungles into new purpose-built fortified villages in order to remove the insurgent support base within the country—and it was very effective (Peng 2003; Barber 2005). Insurgencies sometimes obtain external support from other nations, which allows for a continued growth phase even beyond the resources available in-country (the Soviet Union and China assisted communist insurgencies in Vietnam and elsewhere, and the United States supplied and supported the mujahedeen resistance to the Soviet occupation of Afghanistan). Insurgencies are famous, however, for surviving on minuscule resources. In Vietnam the United States heavily overestimated the Viet Cong's reliance on supplies from North Vietnam and assumed that they could be starved by massive bombing of the Ho Chi Minh trail. Official reports gradually concluded that this had virtually no effect on Viet Cong activity (Pape 1996; Budiansky 2004). Then Secretary of State Robert MacNamara left the White House when the failure of this strategy—his brainchild—became apparent. As Robert Pape concluded, "Guerilla warfare required little in the way of supplies and next to nothing at all from North Vietnam" (Pape 1996, 192). Between 1965 and 1968, for example, the entire 200,000-strong communist force used less than 380 tons of supplies (food, ammunition, medical supplies, etc.) a day. In stark contrast, just one American infantry division consumed 750 to 2000 tons per day (Pape 1996).

Nevertheless, with or without external support, insurgencies at some point become limited by a lack of resources, just as conventional armies do. Vietnamese sources admitted after the war that, although U.S. bombing had failed to break them, their capacity had been drastically reduced and the war effort endangered (Military History Institute of Vietnam 2002). Even superpowers experience density dependence: a million-man army cannot be committed to the front all at once, and fielding large forces overseas is hugely complex and expensive (van Creveld 2004). For example, the U.S. commitment to Iraq amounts to $2 billion a week (Iraq Study Group 2006).

Biological populations are notoriously difficult to eradicate when they occupy large and diverse resource spaces in an ecosystem (Begon et al. 2005). Similarly, ceding territory to insurgent populations allows them to grow to larger sizes. Such territories need not be under the total control of the insurgents, they may simply be anarchic population centers officially under the state's control but nevertheless fertile ground for insurgent growth. In the case of the violent Iraqi conflict, for example, the carrying capacity of insurgents appears to change in a steplike fashion as territory is claimed and ceded by the United States and allied forces. The development of the "green zone" in Baghdad (and, by implication, the vast "red zones" outside it), and the abandonment and subsequent recapture of Falluja by U.S. forces in 2004, have both had dramatic effects on the insurgency's propensity for growth.

Although territory (or space) is a crucial resource for almost all populations, the carrying capacity of insurgent populations can be manifested in different ways. For example, even if government control is equal everywhere, area A may lend greater civilian support to the insurgency than area B. Thus, area A may be more fertile ground for insurgent population growth and elevate the effective carrying capacity. Other factors that boost carrying capacity may include terrain (e.g., 1 km$^2$ of rainforest will allow a higher carrying capacity than the same area of arctic tundra), or geography (e.g., coastal or border areas offer more supply routes and methods to feed the insurgency than would an isolated inland area). In general, insurgencies tend to be more successful where the population is widely distributed in the countryside, where the terrain is conducive to concealment, and where the road or rail infrastructure useful to government forces is poor (as was the case in Vietnam). In general, insurgencies tend to be less successful in deserts or archipelagos, where insurgents cannot easily jump between islands. Rebellious Iraqi villages in the 1920s were singled out and bombed flat by the Royal Air Force, and insurgents in the Philippine and Indonesian islands were easily isolated and trapped (Galula 1964; Fearon and Laitin 2003; Rayburn 2006). For a realistic model of population dynamics, it is essential to account for density dependence.

## Mortality

Populations have both intrinsic and extrinsic processes that decrease population size. In biological populations, intrinsic processes (such as disease or natural mortality) are sometimes as important as extrinsic processes (such as predation, environmental disturbance, or changes in available resources). Among insurgent populations, intrinsic processes are unlikely to have a significant effect because, as noted above, they operate on much longer time scales. Much more important to insurgent populations are extrinsic factors: combat mortality and capture. Although radically different in their nature, these processes have the same effect on a population: they remove individuals from the conflict. If the rate of combat mortality and capture exceeds the rate of recruitment *into* the population (and as long as prisoners are not released), then population size will decrease. Note that emigration *from* the population (due to changes in personal beliefs, the political situation, improved social conditions, etc.) is a recruitment process (albeit, a negative one) and independent of mortality. Altering the environment to incentivize emigration (negative recruitment) offers an indirect solution to insurgency growth and reduces the need for direct conflict.

Aside from combat mortality and capture, disease and accidental deaths are sometimes important phenomena in wartime. For example, disease killed far more soldiers in the Crimean War than did combat (Thayer 2004; Wheelis et al. 2006). As of January 2007, 20% of U.S. deaths in Iraq resulted from "non-combat-related" incidents such as helicopter crashes and other accidents (Iraq Coalition Casualty Count 2007). Data on accidental deaths are not available for insurgent populations, but it is reasonable to assume that proportions are similar (they do not have aircraft to crash in, but they do have malfunctioning improvised explosives, car crashes, and so on). The point here, in any case, is that combat mortality and capture are by far the more important factors.

## Objectives, Methods, and Data

### Objectives

We model the population dynamics of two well-known insurgencies: the Iraq War (2003– ), the ongoing insurgency since the U.S. led invasion in 2003; and the Malayan Emergency (1948–1960), the communist insurgency on the Malaysian Peninsula led by Chin Peng.

Iraq has recently undergone a transition from a single-population insurgency (for which data about insurgency size and mortality are available) to a more complex multipopulation sectarian conflict (for which data are not available). Although there is considerable debate about if or when the Iraq conflict constituted "civil war" (Fearon 2006; Wong 2006), the important point for our analysis is to limit our model to the initial growth period of the insurgency. We therefore use data only up until the February 2006 bombing of the al-Askari mosque in Samarra, an event broadly seen as sparking the onset of major sectarian conflict in Iraq.

By fitting a simple population model to insurgent population data from Iraq, we can: (1) explore the initial growth phase of the insurgent population; (2) project potential population trajectories into a hypothetical future—which assumes the transition into sectarian conflict had been avoided; (3) examine how the manipulation of different population parameters (recruitment, carrying capacity, and mortality rate) are likely to have influenced the survival of the population; and (4) based on these relationships identify policy recommendations for counterinsurgency strategy. The counterinsurgency campaign in Malaya ultimately prevailed (the insurgency was defeated in 1960). Therefore, fitting the same simple population model to both the growth and decay phases of the Malayan insurgency provides the opportunity to estimate what factors may have changed to end the conflict.

## Methods

Following the same methodology that is used to model biological populations that are exposed to external mortality regimes, we sought time-series data on two parameters: insurgent population size and insurgent mortality or capture. We could then use the model to estimate other variables that are much harder to obtain and verify (yet are crucial to monitoring success and developing effective strategies): recruitment rate and the theoretical carrying capacity of the population.

## Data

For Iraq, monthly estimates of the insurgent population size and mortality inflicted by coalition troops were taken from the Brookings Institution's *Iraq Index* (O'Hanlon and Kamp 2006). For Malaya, yearly estimates of the insurgent population size and mortality inflicted by the Malayan Security Forces were taken from United Kingdom Royal Air Force records for the full length of the Malaya Emergency 1948–1960 (Royal Air Force 1970). Estimates of the insurgent population size in Iraq are speculative, and therefore analyses based on these data must be treated with caution. However, the *Iraq Index* serves as a well respected and frequently updated data source that is maintained and used by experts in the field.

## Fitting a Logistic Model with External Mortality

Due to the speculative nature of insurgent population size estimates (especially in Iraq), and the low number of discrete data points in the population size time series, we used the simplest population growth model with density dependence—a discrete time logistic model—to explore the dynamics of insurgent populations (Turchin 2003):

$$\Delta N = r\left(1 - \frac{N}{K}\right)N.$$

In this model, $\Delta N$ is the estimated change in insurgent population size over a single time step (1 month and 1 year for the Iraqi and Malayan conflicts, respectively) and is based on the population size at the start of the time step ($N$), the recruitment rate ($r$), and the population carrying capacity ($K$). When $N$ is small, the expression inside the parentheses is close to 1, and the population can grow freely without constraints. When $N$ approaches $K$ in size, the expression converges on zero, and population growth rate slows down (i.e., growth depends on population density—this is the density-dependence element).

The population size in the next time step $N_{t+1}$ is equal to the current population size $N_t$ plus $\Delta N$. However, if an external force inflicts mortality on the population, this can be taken into account as (Roughgarden 1998):

$$N_{t+1} = N_t + r\left(1 - \frac{N_t}{K}\right)N_t - m_t,$$

where $m_t$ is external mortality at time $t$ inflicted by counterinsurgency forces. That is, the insurgent population size at $t + 1$ is equal to population size at $t$, plus the natural change in population size (immigration and emigration), minus mortality inflicted by external counterinsurgency forces.

This simple logistic model with external mortality was fitted to the data from the Malayan and Iraqi insurgencies. Given the data we did have on $N$ and $m$ (Royal Air Force 1970; O'Hanlon and Kamp 2006), the model used least square optimizations to find the best estimates of the two unknown parameters: carrying capacity $K$ and recruitment rate $r$, which were assumed to remain constant through time. This assumption is unrealistic because rates of recruitment and population carrying capacity would constantly change due to the myriad factors mentioned in the introduction (e.g., changes in socioeconomic conditions or territory under control). Therefore, the model results should be thought of as best estimates given the lack of more detailed data about these parameters.

Our model also assumes that the number of recruits is proportional to the number of fighters. This is a logical assumption but remains to be empirically tested. A related consideration is the mechanism of recruitment. Insurgents may arise spontaneously (because of, say, bad socioeconomic conditions, or as a response to government brutality). Alternatively, active insurgents may recruit new ones through social networks, a phenomenon that would lend itself to epidemiological models (see the appendix and work by Lafferty, Smith, and Madin, this volume). This is an important area for future scrutiny, since different scenarios may lead to different dynamics (Turchin 2003, 117).

Finally, our model assumes that the insurgent population behaves like one large population, whereas, in reality, it is made up of many smaller but integrated populations with differing dynamics. For example, recruitment and resource acquisition are heavily dependent on spatially and socially explicit factors, which may be modeled more accurately using metapopulation- or agent-based models (Macdonald and Johnson 2001; Cederman 2002; Epstein 2006). However, such approaches require much finer scale and more detailed data about subpopulations and their geographic relationships with model parameters; such detail is very hard to come by. Therefore, given the constraints of the data, the single-population model that we present here can be thought of as the average parameter estimates for the insurgency as a whole.

The transition into ethnic conflict in Iraq also reflects the transition from a single-population "insurgent" model to a more complicated multipopulation "sectarian" model in which the role of mortality inflicted by coalition forces becomes less central. Therefore, as mentioned above, we do not model data for the Iraqi insurgent population beyond the February 2006 bombing of the al-Askari mosque. The Iraq model focuses on "hindcasting" (retrospectively analyzing) the growth of the insurgency and forecasting potential scenarios had the transition into sectarian conflict not occurred. These may be particularly useful in determining what would have been required to quell the insurgency *before* the onset of sectarian conflict.

The data for the Malayan insurgency appears to have two distinct phases (an escalation phase and a decline phase) and, therefore, very different model parameter values during each phase. We first fit the model to the growth phase to approximate the insurgency's initial underlying population parameters and for a more appropriate comparison with the Iraqi insurgency, and we subsequently fit the model to the decline phase to give insight as to the changes in population parameters that led to population decay.

To identify the optimal point at which to divide the Malaya time series, we fitted the population model to a progressively greater number of time points in the series (out of a total of 13). The model used was the one fitted to the first 5 points (i.e., 1948–1952), selected on the basis that using more or fewer years produced estimates of $r$ and $K$ that were clearly unrealistically high or low. Interestingly, however, 1952 turns out to be a historically meaningful turning point, because this was when counterinsurgency efforts came under the military command of General Gerald Templer (following the assassination of the British high commissioner, Sir Henry Gurney). By all accounts, Templer was a force of nature who comprehensively and effectively revolutionized military, social, and political efforts to combat the insurgency (Nagl 2002; Peng 2003).

## Model Scenarios

Once population parameters $K$ and $r$ were estimated by fitting the models, future trajectories for the insurgent population could be calculated. These could be trajectories assuming all basic parameters stayed the same, or, more interestingly, they could be alternative scenarios in which model parameters were altered (corresponding to hypothetical changes in circumstances or counterinsurgency strategy). For each conflict, extrinsic mortality $m$, carrying capacity $K$, and recruitment $r$ were varied in turn (one at a time) to illustrate their potential effects on the insurgent population size. More complicated scenarios are possible in which multiple parameters could be altered at once, or in which parameters change as a function of time. However, given the uncertainty of the data, the low number of data points, and the simplicity of the population model, more complex scenarios were avoided in the present study (we judged they would be too speculative).

Two additional analyses were conducted. First, a significant linear relationship was detected between the mortality inflicted on the Iraqi insurgent population and time, indicating a gradual, although highly variable, increase in insurgent mortality over the course of the conflict (basically, coalition forces have killed or captured an increasing number of insurgents). Because of this, we ran an additional future scenario for Iraq in which mortality continues to increase as a function of time at the same rate as it has done for the first 35 months of conflict (according to a best-fit linear model). Second, for the Malaya conflict, data on $N$ and $m$ are available for the additional years of conflict 1952–1960—beyond the growth phase during 1948–1952 to which the model was initially fit. This presents the opportunity to fit the population model to the remaining time series (using the 1952–1960 estimates of $N$ and $m$) to give best estimates of carrying capacity $K$ and recruitment rates $r$ during the decline and ultimate defeat of the Malayan insurgency.

## Results

### Estimates of Carrying Capacity and Recruitment Rates

*Iraq* The best-fit model for the Iraqi insurgency (Fig. 11.1, solid line) suggested a carrying capacity for this population of approximately 23,000 insurgents, and a recruitment rate of approximately 0.6 which, in the absence of density dependent processes, corresponds to a monthly influx equal to 60% of the population. An estimate of the insurgency's population growth given no external mortality (Fig. 11.1, dotted line) suggests that mortality inflicted by coalition forces slowed the initial growth of the

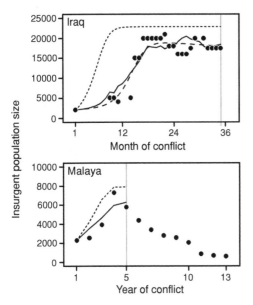

FIGURE 11.1 Logistic population model with external mortality fitted to population size (black dots) and mortality estimates of the ongoing Iraqi insurgency (2003–2006) and the escalatory phase of the 1948–1960 Malaya insurgency (solid lines). Dotted lines illustrate the hypothetical growth of insurgencies given the carrying capacity and recruitment estimates from the model but assuming no external mortality. Dashed line in the Iraq panel shows the model fit when extrinsic mortality is modeled according to a linear regression function describing the empirical increase in mortality over time (see text for details).

insurgent population, but was not enough to stop the population reaching a size similar to the model carrying capacity. Finally, the dashed line (Fig. 11.1) illustrates the best-fit model when mortality inflicted by coalition forces increases linearly through time (as has apparently occurred according to the linear regression of insurgent mortality data over time).

*Malaya* The best-fit model for the first five years of the Malayan insurgency (Fig. 11.1, solid line) suggested a carrying capacity for this population of approximately 7900 insurgents, and a recruitment rate of approximately 1.2, which, in the absence of density dependent processes, corresponds to a more than doubling of the population from year to year. Remarkably, this recruitment rate is approximately eight times lower per unit time than for the Iraqi insurgency (for Iraq, $r = 0.6$ [monthly], for Malaya $r = 1.2$ [yearly], which is equivalent to a monthly rate of $r = 0.07$; note that this is not simply 1.2 divided by 12, because one has to account for cumulative gains per month). An estimate of the insurgency's population growth given no external mortality (Fig. 11.1, dotted line) suggests—identically to the Iraq case—that mortality inflicted by Malayan Security Forces slowed the initial growth of the insurgent population but, again, was not enough to stop the population reaching a size similar to the model carrying capacity. There was not enough data in this case to produce a model accounting for any possible variation in mortality over time.

## Iraq: Model Scenarios

*Mortality* In all Iraq analyses, "mortality" includes insurgents killed or captured. Assuming that the transition into sectarian conflict had been avoided and the above estimates of carrying capacity and recruitment rate for the Iraqi insurgency remain the same for the remainder of the conflict (after January 2006, month 35 of conflict), future projections of insurgent population size can be estimated for scenarios in which mortality inflicted by the coalition forces varied (Fig. 11.2, top panel). For example, if coalition forces ceased offensive operations entirely, the insurgent population is predicted to have rapidly increased in size toward the carrying capacity and reach it in about 6 months. (Note that such a policy could also lead to a subsequent increase in $K$ because, for example, more safe areas would become available. Conversely, $K$ could subsequently decrease if the presence of the counterinsurgency campaign itself is a major resource base for insurgents).

If mortality remains similar to estimates for January 2006 (approximately 2570 insurgents killed or captured per month), then the population size is predicted to remain indefinitely at approximately 17,500 insurgents. By increasing mortality by 50% above the January 2006 estimates (i.e., to approximately 3900 per month), the insurgency is predicted to collapse within approximately 20 months. A 100% increase in mortality predicts a collapse within 10 months. As we mentioned earlier, insurgent mortality rates in Iraq have actually increased over time. If we assume that this increase continues along the same linear trajectory, the model predicts that it would take 55 months (or about 4.5 years) for the coalition forces to neutralize the insurgency (Fig. 11.2, top panel, dashed line).

*Carrying Capacity* Assuming that the levels of mortality inflicted by allied forces remains the same as in January 2006 (approximately 2570 casualties per month) and the estimated recruitment rate remains at 0.6, we also projected the future insurgent population size for alternative carrying capacities (Fig. 11.2, middle panel). If enough territory or other resources were ceded to increase the carrying capacity to 30,000, the population is estimated to increase in size to approximately 25,000 insurgents. If the carrying capacity were reduced by half, the insurgent population is estimated to become extinct in approximately 10 months. Carrying capacities of 16,500 and 17,000 represent the boundaries of an unstable equilibrium, above which the population is sustained for a long time into the future, and below

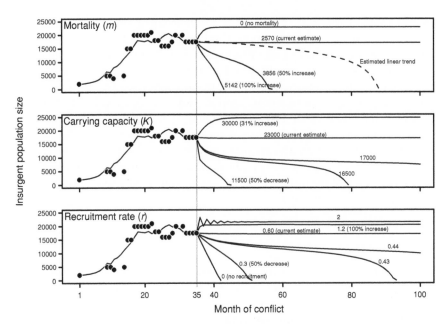

FIGURE 11.2 Projected future trajectories for the Iraqi insurgency (after January 2006, or month 35 of the conflict) calculated by altering (1) extrinsic mortality *(top panel)*, (2) carrying capacity *(middle panel)*, and (3) recruitment rate *(bottom panel)*. Black dots represent actual insurgent population size. Numbers labeling the curves refer to alternative values of each parameter. Based on the fitted logistic model, extrinsic mortality is 2570 per month, carrying capacity is 23,000, and recruitment rate is 0.60, unless otherwise stated. The dashed line in the "Mortality" panel illustrates the estimated population trajectory if monthly mortality increases in the same linear fashion as it has over the first 35 months of the conflict (see text).

which the population tends to extinction. Since the real population size is thought to be above this value, coalition forces would have had to actively reduce the available resources in order to stem the insurgency via carrying capacity alone.

*Recruitment Rate* Assuming that the January 2006 level of mortality remains the same (approximately 2570 casualties per month) and the estimated carrying capacity of the population remains at 23,000, we projected the future insurgent population size with alternative recruitment rates (Fig. 11.2, bottom panel). Even doubling the recruitment rate does not increase the insurgent population substantially because the carrying capacity creates a ceiling close to the existing population size. However, note that potential increases in recruitment rate are likely to co-occur with increases in the population's carrying capacity, so it could push this ceiling higher (for example, an increase in recruits may result from or cause an increase in sympathizers or safe houses). Interestingly, if recruitment rate becomes too high, the population becomes unstable because limiting resources become overused (oscillating line in Fig. 11.2, bottom panel). A 50% reduction in recruitment rate under the current mortality regime would effectively neutralize the insurgent population in approximately 15 months. The model indicates that it would still take approximately 7 months for coalition forces to neutralize the insurgency even if recruitment into the population was stopped altogether. As with carrying capacity, there is an unstable equilibrium between $r = 0.43$ and $0.44$, above which the population persists, and below which it will become extinct (again, assuming January 2006 levels of mortality and carrying capacity).

## Malaya: Model Scenarios

The fate of the Malayan insurgency is well known: it was defeated in 1960. However, by fitting our model to the decline phase of this conflict, we can: (1) examine different scenarios of how the insurgent population might have fared if things had gone differently; (2) see whether changes in specific population parameters match known historical changes in policy or strategy, which would validate the credibility of these kinds of models; and (3) derive the likely population parameters that changed to end the conflict.

*Mortality* Assuming recruitment and carrying capacity remained the same following the growth phase of the insurgent population (i.e., $r = 1.2$ and $K = 7900$), the "archived" (i.e., as known from the historical data) levels of yearly external mortality would not have quelled the insurgency (Fig. 11.3, top panel). Several other scenarios are given for different hypothetical mortality rates, illustrating that had mortality increased to 5000 per year, the insurgency could have died out as early as 1954.

*Carrying Capacity* Assuming the archived levels of mortality and that recruitment rate remained the same following the growth period ($r = 1.2$), we also projected the future insurgent population size given different carrying capacities (Fig. 11.3, middle panel). If the carrying capacity were reduced by 50%, the insurgent population would still have persisted. Carrying capacity would have had to be reduced to approximately 1000 or less to cause a collapse (which seems a very small figure, but remember this assumes that all other parameters stayed the same, and the history of the conflict suggests that they did not).

*Recruitment Rate* Assuming archived mortality and that the carrying capacity remained constant following the growth period ($K = 7900$), even halving the recruitment rate would have only briefly dented the population size

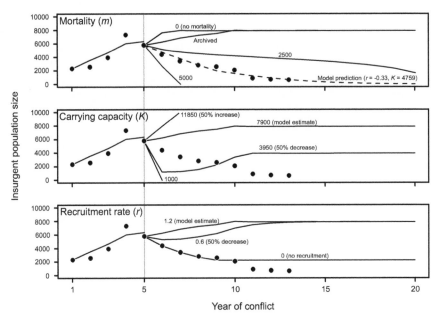

FIGURE 11.3 Projected trajectories for the Malaya insurgency following the growth
phase (1948–1952, or years 1 to 5) calculated by altering (1) extrinsic mortality
*(top panel)*, (2) carrying capacity *(middle panel)*, and (3) recruitment rate *(bottom
panel)*. Black dots represent actual insurgent population size. Numbers labeling
the curves refer to alternative values of each parameter. Based on the fitted logistic
model, carrying capacity is 7900, and recruitment rate is 1.2, unless otherwise
stated. Mortality per year is as "archived" (i.e., as known from the historical data)
where labeled in the first panel and for all curves in the second and third panels.
The dashed line in the "Mortality" panel illustrates the best fit of the logistic
model with archived extrinsic mortality (see text).

(Fig. 11.3, bottom panel), which would have grown back to similar levels in
five or six years due to declines in archived mortality. Note that the population persists at a low level even if $r = 0$ (i.e., no recruitment at all), because
historical mortality rates decline to zero (as the insurgency subsided), while
lack of recruitment means that the population cannot recover to its carrying
capacity. How then could there have been an actual drop in population size
on the ground, as we can see from the historical data? Using the archived
mortality estimates *and allowing recruitment and carrying capacity to vary*, a
best-fit model (Fig. 11.3, top panel, dashed line) suggests that during the
period of decline, insurgent recruitment rate was approximately –0.33 (i.e.,
the population was decreasing by a third each year via emigration) and carrying capacity was around 4800. The history of the conflict ties in very
nicely with this result. Toward the latter years of the conflict, British and

Malayan forces were successful in encouraging defectors or captives to return to the jungle and convince their erstwhile comrades to defect as well. Significant financial rewards were offered for defecting, as well as for bringing out other defectors (Nagl 2002; Barber 2005). In this way, the counterinsurgency forces were effectively able to make the recruitment rate negative. Defectors begat defectors and the insurgent population spiraled down, while the widening gaps in their ranks further hampered their effectiveness and persistence (cf. the Allee effect).

## Discussion

In both Iraq and Malaya, there appear to have been critical moments at the start of the insurgencies when the opportunity to defeat them was greatest. In Iraq, coalition forces seem to have made a significant dent in the potential insurgent population size in the initial months of the conflict (compare dotted and solid lines in Fig. 11.1). However, the opportunity to stamp out the insurgency may have long passed even by that time. Before the invasion of Iraq, the U.S. 4th Infantry Division was scheduled to open up a third front from the north, but at the last minute the Turkish government declined authorization for U.S. forces to pass through its territory. Although this clearly had little effect on the success of the invasion itself, it is likely that it had significant ramifications for the insurgency to come. Coalition troops attacking from the north would have passed directly through the "Sunni Triangle," the insurgent stronghold that was instead left to fester and to become the most violent region after the end of major combat. Moreover, two Iraqi Republican Guard divisions vanished in this area toward the end of the U.S. invasion, and many of those men (and their weapons) no doubt contributed to the ensuing insurgency. In addition to this, for several days immediately after the fall of Baghdad, almost nothing was done by U.S. forces to control the civilian population, provide security, prevent looting, or impose law and order. Finally, the entire Iraqi army was disbanded—several thousand trained, armed, and disenfranchised men were suddenly unemployed. All this was fertile ground for an insurgency to gain a foothold (Clark 2003; Fallows 2004, 2005; Johnson 2004; Woodward 2005). Certainly, the models suggest that the initial days following regime change in future analogous conflicts will be crucial. The models also point to policies and strategies that might have increased the probability of defeating the insurgency *before* Iraq descended into major sectarian violence.

Coalition forces could have intensified their efforts on inflicting mortality on the insurgent population. Our simplistic model suggests that by removing 50% more insurgents per month, and ensuring that the territory occupied by insurgents and recruitment into the population would not change too much, the insurgent population could have been defeated in

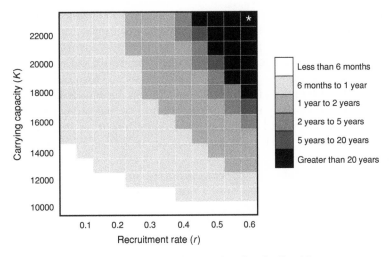

FIGURE 11.4 Estimated times until extinction for the Iraqi insurgent population given alternative recruitment rates and/or population carrying capacity, assuming that extrinsic mortality remains at the January 2006 level. An asterisk indicates the January 2006 estimates of $r$ and $K$.

under two years (just before the presidential election in 2008). However, constraints on the deployment and combat activities of U.S. and coalition forces inhibited any long-term efforts to ramp up offensive operations (Iraq Study Group 2006). Furthermore, efforts to increase insurgent casualties were likely to cause an increase in U.S. casualties as well, which, without clear signs of impending victory, undermines domestic support (Gelpi et al. 2005; Johnson and Tierney 2006). In any case, capturing and killing insurgents may not be an effective long-term counterinsurgency strategy, compared with the more important targets of civilian "hearts and minds" (Galula 1964; Beckett 2001; Aylwin-Foster 2005).

As long as coalition forces maintained their gradual increase in inflicting mortality, the insurgent population was predicted to decline, if slowly (and assuming all else remained equal). However, even this minimum objective was unlikely to be achieved with the prospect of U.S. troop withdrawals, and with the transfer of counterinsurgency responsibilities to Iraqi security forces, which proved much less effective (Fallows 2005; Biddle 2006). Given ethical, political, and practical constraints, therefore, the models are particularly interesting in highlighting what other counterinsurgency strategies are open to coalition forces that may prove to be as, or perhaps even more, effective.

Moderate reductions in recruitment or carrying capacity led to a fairly significant decline in insurgent population size—indeed, greater declines than resulted from equivalent changes in mortality rates (see Fig. 11.2). Therefore, recruitment and the insurgents' resource base may be key targets for manipulation. Tangible improvements in the political, economic, and social situation in Iraq may cause recruitment rates into the insurgency to fall (though perhaps less effectively than in past communist counterinsurgencies such as in Malaya or Vietnam [Biddle 2006]). This goal is likely to have a double whammy effect: improved conditions would also decrease the effective carrying capacity of the insurgent population as government institutions, civilians, and coalition forces reinforce their influence over insurgent-prone territories (in turn making their task of defeating the insurgency easier).

Figure 11.4 illustrates the effect of different recruitment rates or carrying capacity on the predicted time it would have taken coalition forces to defeat the insurgency (assuming, as before, that extrinsic mortality remained similar to estimates for the period up to January 2006, before major sectarian violence began). If coalition forces did not alter the way they were fighting the insurgency in January 2006, the insurgent population is predicted to continue for more than 20 years into the future (if this sounds overly pessimistic, think of Vietnam, or even Malaya which, though ultimately successful, took 12 years). However, by changing the carrying capacity or recruitment rate of the insurgent population, the expected time until the end of conflict can be significantly diminished. We saw in Figure 11.2 that halving recruitment or carrying capacity could lead to the defeat of the insurgency in about a year. In combination, this effect would be magnified, such that even moderate changes—if they impact both parameters at once—can lead to a relatively quick victory by counterinsurgency forces. For example, Figure 11.4 shows that if $K$ and $r$ were reduced by just a third each ($K \sim 15,000$ and $r \sim 0.4$), the insurgency is predicted to collapse in 6 to 12 months. Add some extra effort in capturing key leaders, say, and it could be over even sooner or, in another scenario, defeated in 12 months even if recruitment rates refused to budge. Moreover, as suggested above, it is likely that these parameters are linked, such that a decrease in one would cause a decrease in the other. Clearly, a combination of strategies targeting recruitment, carrying capacity, and mortality would be most likely to lead to a coalition victory. The models strongly suggest that all of them are critical, and therefore the coalition should not rely solely on the dangerous and politically costly task of killing and capturing insurgents.

## Model Lessons: Comparing Malaya and Iraq

Since 1945, civil wars—of which insurgencies are one type—have an average duration of over 10 years (Fearon 2006). Even successful counterinsurgencies, such as that in Malaya, can take 12 years. It may therefore be far too early to judge success or failure in Iraq. Nevertheless, important differences may explain the diverging trends during the early stages of both conflicts. Although Malaya and Iraq are distinguished by many unique factors, comparisons can shed light on common themes of success and failure, and of the major obstacles in each case. There is a considerable literature on the lessons of different counterinsurgency campaigns (e.g., Nixon 1985; McNamara 1996; Woodruff 1999; Beckett 2001; Gilbert 2002; Nagl 2002; Walton 2002; Kramer 2004). Here, we offer some novel insights by limiting our comparisons to the key parameters of our models. Our conclusions do not bode well for the counterinsurgency campaign in Iraq—especially now the task has been made harder with the increasing sectarian conflict.

Mortality was a key variable in both conflicts. In Malaya, however, counterinsurgency forces enjoyed some battlefield advantages. First, Malayan insurgents were (for the most part) an easily identifiable ethnic group (Chinese), who often also wore telltale uniforms. Iraqi insurgents generally look and dress like everyone else. Second, Malayan insurgents tended to confine themselves to the jungle—a difficult environment in which to track them down, of course, but at least counterinsurgency forces knew the geographical subset of areas in which they could search for insurgents and their camps. Iraqi insurgents blend in with the crowds in urban environments that are often too dangerous for counterinsurgency or police forces to patrol (without significant armed support). Third, the intensity of the military effort remained high in Malaya, with a mixture of British conscripts, regular troops, significant police participation, and indigenous Malayan forces, all of which benefited from domestic support at home. In Iraq, U.S. troops were too few to begin with, have little likelihood of long-term reinforcement, and receive limited and largely ineffective help from indigenous forces, and there is increasing antiwar sentiment among the U.S. public and Congress (Americans support the troops, but they do not like them being in Iraq), as well as internationally (troops have been withdrawn by numerous countries that made up the initial coalition).

Carrying capacity also varied between the two conflicts. First, the Malayan communists did not have a great amount of support from outside powers, and the only land border on the Malaysian peninsula was the narrow northern isthmus connecting with Thailand (and Thais were neither ethnically Chinese nor communist). Iraqi insurgents enjoy a vast and porous border with numerous Sunni and Shi'a countries with significant populations motivated to support them, passively or actively, or even join them (estimates of foreign fighters in Iraq vary, but they account for at least a significant minority). Second, Malayan communists relied for logistical support on the many thousands of ethnic Chinese in the countryside. The Briggs Plan took the extraordinary measure of uprooting 400,000 of these people and moving them into purpose-built guarded villages. In Iraq, there are regions in which even armored U.S. troops cannot go, let alone entertain the concept of sealing off a specific identifiable subset of insurgent sympathizers.

Recruitment rates also had very different dynamics in the two conflicts. First, the pool of potential recruits in Malaya was somewhat limited: mainly ethnic Chinese, committed communists, or others dissatisfied with the colonial regime. Moreover, the move toward Malayan independence was already well underway for the existing government in Kuala Lumpur, so dissatisfied colonial subjects had only to wait for the scheduled turnover in 1957 rather than join an insurgency. In Iraq, the pool of potential recruits includes not only a wide variety of dissatisfied Iraqis, including Sunni and Shi'a militants, but also anyone opposed to the coalition or the new Iraqi government, from across the global Muslim population. Second, as the Malaya conflict progressed and independence neared, there was less and less reason to join the insurgency. As the Iraqi conflict goes on, U.S. troops remain, the chaos persists, the Iraqi government proves ineffective, and sectarian violence increases, there are arguably more and more reasons to join the insurgency. Third, there was no religious component to the Malayan insurgency; it was purely a political movement. As the political situation changed, therefore, so ineluctably did the rationale for insurgents to fight. In Iraq, although many insurgents may not be religiously motivated, the presence of Christian occupying troops on Islamic lands is a significant rationale for many insurgents to fight, even die, to some extent irrespective of the political situation. Lastly, the counterinsurgency forces in Malaya were able to offer financial rewards to buy insurgents out of the jungle and pay them to bring others out too, turning the recruitment rate negative. Such a method would be extremely unlikely to work in Iraq. Our models suggest the monthly recruitment rate in Iraq is eight times the rate during the growth phase in Malaya. With such a significant difference in this crucial parameter, even innovative and successful strategies to recruit insurgents out of the existing population with financial or other incentives may fail to defeat the burgeoning insurgency in Iraq.

To conclude, all models are simplifications of reality; however, that is precisely why they are useful. Only by distilling the complexity of messy events down to key parameters can we hope to identify patterns amidst the noise—patterns that in this case may reveal counterinsurgency strategies that will succeed and others that will fail. Numerous political, social, and economic factors obviously influence the dynamics of conflict, but at the end of the day, an insurgency is a group of individuals that can either grow or decline. If $m$ is small and $r$, $K$, and $N$ are large, the insurgents will win. If $m$ is large and $r$, $K$, and $N$ are small, the insurgents will fail. These are the magic numbers that a counterinsurgency campaign must manipulate:

simply matching or outnumbering raw numbers of troops on the ground is not the answer. Russian troops and police have been unable to defeat the rebels in Chechnya even though they outnumber them by more than 50 to 1 (Kramer 2004), and according to David Galula, the French could not have defeated the insurgency in Vietnam "even if they had been led by Napoleon" (Galula 1964, 32). As in Malaya, success in Iraq will depend on novel strategies and novel approaches.

Although ecological approaches demand further development, they offer a number of novel insights for understanding how insurgencies change over time, and what strategies are most likely to defeat them. Seeing insurgencies through an ecological lens may also crystallize some important underlying mechanisms. For example, in biological populations, *adaptation* is the key to population survival or extinction. In his comparison of counterinsurgency campaigns in Malaya and Vietnam, John Nagl argued that the difference in outcome was not to do with experience—both nations had plenty of experience with occupations. The difference was that the smaller British army was able to learn from its mistakes in Malaya and adapt accordingly, while the U.S. army in Vietnam was not (Nagl 2002). Some of these same differences between the British and U.S. approach to insurgency have attracted attention in Iraq as well (Aylwin-Foster 2005; T. Taylor, pers. comm., 2005). The decisive factor in the victory or defeat of counterinsurgency operations, therefore, may be the ability to achieve efficient adaptation, a concept that leads directly back to Darwin.

## Appendix: Alternative Models and Future Opportunities

Future studies could extend and improve our approach in a number of ways. For example, one could build models with new case studies (e.g., Vietnam or Chechnya), better data (more reliable, more frequent or longer term data), additional parameters (such as the number of counterinsurgency troops), or alternative model formulations. We suggest some of the main alternative models that one could use:

*Competition models.* If it can be assumed that insurgent and counterinsurgent populations compete for the same resource (such as territory, or the "hearts and minds" of civilians) competition models offer an alternative for understanding the dynamics of two (or more) interacting populations, and the propensity for any population to either thrive or become extinct. However, competition models typically focus on population sizes through time and do not incorporate mortality inflicted by one population on the other explicitly, although this parameter can be added. Additionally, the use of competition models in scenarios where one population is superior in size and strength is questionable, because different resource bases are likely to be the cause of these disparities.

*Predator-prey models.* Predator-prey models are a less likely candidate for modeling insurgency and counterinsurgency populations because they describe directed conflicts. The basic model assumes that the prey population utilizes its own independent resource, and the prey population is itself the resource for the predator population. Therefore, a decrease in the prey population (typically by predator consumption) leads to a decrease in the size that the predator population can sustain, which allows for the prey population to grow back if it has not already been depleted. One scenario in which this model might be effectively applied is in gauging the response of counterinsurgency troop levels to decreases in insurgent population size (after all, troop levels are usually altered to some extent, or deployed differently, depending on the intensity of the insurgency). The predator-prey model has been used in some studies of unbalanced conflict by holding predator population size constant (Turchin 2003, 101).

*Epidemiological (disease) models.* These are explored elsewhere in this volume (Lafferty, Smith, and Madin) and offer another approach with its own pros and cons. Such models are good at identifying thresholds and have a useful categorization of susceptible, infected, and resistant subpopulations that might be very relevant to some types of ideological insurgencies (see Stares and Yacoubian 2005).

*Metapopulation models.* An insurgency might behave more like a collection of somewhat independent subpopulations with different properties and behavior than a "normal" single population. For instance, a metapopulation model might lead to insights about the importance of insurgents moving between focal areas of activity. Spatial heterogeneity in conditions and parameters could be integrated into such a model. Predictions might be quite different (e.g., quell the insurgency in Falluja, and Baghdad will collapse on its own). Data collection would have to be on a much finer scale, however.

### ACKNOWLEDGMENTS

We thank Rafe Sagarin and Terrence Taylor for their ideas, advice, criticism, and the invitation to join the working group on Ecological and Evolutionary Models for Homeland Security Strategy, and the National Center for Ecological Analysis and Synthesis for hosting us. Dominic Johnson is indebted to the Branco Weiss Society in Science Fellowship, the International Institute at UCLA, and the Society of Fellows and the Woodrow Wilson School of Public and International Affairs at Princeton University. We also owe our thanks to Scott Field, Simon Levin, Jason Lyall, Peter Turchin, and all the members of the working group for excellent comments and criticisms on the manuscript.

## REFERENCES

Aylwin-Foster, N. 2005. Changing the army for counterinsurgency operations. *Military Review* 85: 2–15.

Barber, N. 2005. *War of the running dogs: How Malaya defeated the communist guerillas, 1948–1960.* London: Cassell.

Beckett, I. F. W. 2001. *Modern insurgencies and counter-insurgencies: Guerrillas and their opponents since 1750.* New York: Routledge.

Begon, M., C. R. Townsend, and J. L. Harper. 2005. *Ecology: From individuals to ecosystems.* Oxford: Blackwell Scientific.

Biddle, S. 2006. Seeing Baghdad, thinking Saigon. *Foreign Affairs* March/April, vol. 85, No 2, 2–14.

Budiansky, S. 2004. *Air power: The men, machines, and ideas that revolutionized war, from Kitty Hawk to Gulf War 2.* New York: Viking.

Cederman, L.-E. 1997. *Emergent actors in world politics: how states and nations develop and dissolve.* Princeton, NJ: Princeton University Press.

Cederman, L.-E. 2002. Endogenizing geopolitical boundaries with agent-based modeling. *Proceedings of the National Academy of Sciences* 99: 7296–7303.

Clark, W. K. 2003. *Winning modern wars: Iraq, terrorism, and the American empire.* New York: PublicAffairs.

Clarke, W. S., and J. I. Herbst, eds. 1997. *Learning from Somalia: The lessons of armed humanitarian intervention.* Boulder, CO: Westview Press.

Ehrlich, P., and S. Levin. 2005. The evolution of norms. *PLoS Biology* 3: e194.

Epstein, J. M. 1997. *Nonlinear dynamics, mathematical biology and social science.* Reading, MA: Addison Wesley Publishing.

Epstein, J. M. 2006. *Generative social science: Studies in agent-based computational modeling.* Princeton, NJ: Princeton University Press.

Fallows, J. 2004. Blind into Baghdad. *Atlantic Monthly.* January/February, vol. 293, No 1, 53–74.

Fallows, J. 2005. Why Iraq has no army. *Atlantic Monthly.* December, vol. 296, No 5, 296: 60–77.

Fearon, J. D. 2006. Iraq: Democracy or civil war? *U.S. House of Representatives Committee on Government Reform, Subcommittee on National Security, Emerging Threats, and International Relations.* 15 September 2006. Available at: http://cisac.stanford.edu/publications/iraq_democracy_or_civil_war/

Fearon, J. D., and D. D. Laitin. 2003. Ethnicity, insurgency, and civil war. *American Political Science Review* 97: 75–90.

Francisco, R. A. 1996. Coercion and protest: An empirical test in two democratic states. *American Journal of Political Science* 40: 1179–1204.

Francisco, R. A. 2004. The dictator's dilemma. In *Repression and mobilization,* ed. C. Davenport, H. Johnston, and C. Mueller). Minneapolis, MN: University of Minnesota Press.

Galula, D. 1964. *Counter-insurgency warfare: Theory and practice.* New York: Praeger.

Gelpi, C., P. D. Feaver, and J. Reifler. 2005. Success matters: Casualty sensitivity and the war in Iraq. *International Security* 30: 7–46.

Gilbert, M. J. 2002. *Why the North won the Vietnam War.* New York: Palgrave.

Iraq Coalition Casualty Count. 2007. Available at www.icasualties.org.

Iraq Study Group. 2006. *The Iraq Study Group report: The way forward—A new approach.* New York: Vintage.

Johnson, D. D. P. 2004. *Overconfidence and war: The havoc and glory of positive illusions.* Cambridge, MA: Harvard University Press.

Johnson, D. D. P., and D. R. Tierney. 2006. *Failing to win: Perceptions of victory and defeat in international politics.* Cambridge, MA: Harvard University Press.

Kramer, M. 2004. The perils of counterinsurgency: Russia's war in Chechnya. *International Security* 29: 5–63.

Macdonald, D. W., and D. D. P. Johnson. 2001. Dispersal in theory and practice: Consequences for conservation biology. In *Causes, consequences and mechanisms of dispersal at the individual, population and community level,* ed. J. Clobert, J. D. Nichols, E. Danchin, and A. Dhondt, 361–374. Oxford: Oxford University Press.

McNamara, R. S. 1996. *In retrospect: The tragedy and lessons of Vietnam.* New York: Vintage Books.

Military History Institute of Vietnam. 2002. *Victory in Vietnam: The official history of the People's Army of Vietnam, 1954–1975.* Lawrence: Kansas University Press.

Nagl, J. A. 2002. *Learning to eat soup with a knife: Counterinsurgency lessons from Malaya and Vietnam.* Chicago: Chicago University Press.

Nixon, R. 1985. *No more Vietnams.* New York: Arbor House.

O'Hanlon, M. E., and N. Kamp. 2006. *Iraq index: Tracking variables of reconstruction and security in post-Saddam Iraq.* Washington DC: Brookings Institution.

Pape, R. A. 1996. *Bombing to win: Air power and coercion in war.* Ithaca, NY: Cornell University Press.

Pape, R. A. 2005. *Dying to win: The strategic logic of suicide terrorism.* New York: Random House.

Peng, C. 2003. *Alias Chin Peng: My side of history.* Singapore: Media Masters.

Rayburn, J. 2006. The last exit from Iraq. *Foreign Affairs* 85: 29–40.

Richardson, L. F. 1960. *Statistics of deadly quarrels.* Pacific Grove, CA: Boxwood Press.

Richardson, L. F. 1978. *Arms and insecurity: A mathematical study of the causes and origins of war.* Pacific Grove, CA: Boxwood Press.

Roughgarden, J. 1998. *Primer of ecological theory.* Upper Saddle River, NJ: Prentice-Hall.

Royal Air Force. 1970. *The Malayan Emergency, 1948–1960.* National Archives, London, AIR 10/8584: Royal Air Force.

Rumsfeld, D. 2003. Global war on terrorism (memo). Available at www.usatoday.com/news/washington/executive/rumsfeld-memo.htm.

Stares, P., and M. Yacoubian. 2005. Terrorism as virus. *Washington Post,* August 23, A15.

Stephens, P., W. Sutherland, and R. Freckleton. 1999. What is the Allee effect? *Oikos* 87: 185–190.

Stern, J. 2003. *Terror in the name of God: Why religious militants kill.* New York: HarperCollins.

Thayer, B. A. 2004. *Darwin and international relations: On the evolutionary origins of war and ethnic conflict.* Lexington: University Press of Kentucky.

Toft, M. D. 2002. Differential demographic growth in multinational states: Israel's two-front war. *Journal of International Affairs* 56: 71–94.

POPULATION MODELS AND COUNTERINSURGENCY      185

Turchin, P. 2003. *Complex population dynamics: A theoretical/empirical synthesis.* Princeton, NJ: Princeton University Press.

Turchin, P. 2005. *War and peace and war: The life cycles of imperial nations.* New York: Pi Press.

van Creveld, M. 2004. *Supplying war: Logistics from Wallenstein to Patton.* Cambridge: Cambridge University Press.

Walton, C.D. 2002. *The myth of inevitable U.S. defeat in Vietnam.* Portland, OR: Frank Cass International Specialized Book Service.

Wheelis, M., L. Rózsa, and M. Dando, ed. 2006. *Deadly cultures: Biological weapons since 1945.* Cambridge, MA: Harvard University Press.

Wong, E. 2006. A matter of definition: What makes a civil war, and who declares it so? *New York Times,* November 26.

Woodruff, M.W. 1999. *Unheralded victory: Who won the Vietnam War?* London: HarperCollins.

Woodward, B. 2005. *State of denial: Bush at war, part III.* New York: Simon and Schuster.

Wright, Q. 1983. *A study of war.* Chicago: University of Chicago Press.

Chapter 12

# THE INFECTIOUSNESS OF TERRORIST IDEOLOGY

Insights from Ecology and Epidemiology

KEVIN D. LAFFERTY, KATHERINE F. SMITH, AND ELIZABETH M.P. MADIN

Terrorism in the twenty-first century is unconventional, unpredictable, and potentially unavoidable. In part, this is because contemporary terrorists are increasingly transnational, industrious, unorthodox in their methods, and decentralized (e.g., Ariza 2006; Ehrlich and Levin 2005). Some have proposed that we view terrorism through the lens of epidemiology, where terrorist ideology is analogous to an infectious agent of threat to global public health (in particular, Stares and Yacoubian 2005). While the terrorist ideology–infectious agent analogy has obvious utility, we recognize that is it is both young and imperfect. Here, we take the next step and investigate the value of this analogy.

Similarities between terrorism and pathogen outbreaks have resulted in parallel, but independent, tracking, prevention, and control efforts. The National Strategy for Homeland Security, for instance, "recognizes that the capabilities and laws we rely upon to defend the U.S.A. against terrorism are closely linked to those which we rely upon to deal with non-terrorist phenomena such as disease" (Office of Homeland Security 2002, 4). If sufficient parallels exist between infectious agents and terrorism, counterterrorist efforts may be able to draw on the already substantial body of theory developed for public health (see the box "Making the Metaphor Practical"). In this chapter, we explore if and how epidemiological theory can increase our understanding of the dynamics and spread of terrorist ideology—the belief structures on which acts of terror are supported, justified, and carried out. We use the terms infectious agent, disease, parasite, and pathogen interchangeably throughout this chapter and as analogs to terrorist ideology. We use our findings to identify key unanswered questions and avenues for future research. In considering this analogy, we do not promote any particular counterterrorist strategy or recommend policy.

## Making the Metaphor Practical

In their informative article "Unconventional Approaches to an Unconventional Threat: A Counter-Epidemic Strategy," Stares and Yacoubian (2005) point out what they perceive to be the three most practical applications of epidemiology and public health to combating terrorism. (They specify Islamist militancy, while we consider terrorism generically.)

1. Epidemiologists observe rigorous standards of inquiry and analysis to understand the derivation, dynamics, and propagation of infectious agents. They seek clarity on the origins, geographical, and social contours of an outbreak: where is the pathogen concentrated, how it is transmitted, who is most susceptible to infection, and why are some immune? Applying the same methodological approach to mapping and understanding terrorism can yield immediately useful guidance on where and how to counter it.
2. Epidemiologists recognize that infectious agents emerge and evolve as a result of complex interactive processes between hosts, pathogens, and the environment in which they live. To make sense of this complexity, epidemiologists deconstruct the key constituent elements of an infectious agent. This model helps to understand the phenomenon in its entirety and anticipate how it might evolve in the future. The same systemic conception of infectious agents can be adapted to understand the constituent elements of terrorism and their evolution (Fig. 12.1).
3. Epidemiologists view infectious agents as complex, multifaceted phenomena. Public health officials have thus recognized that success in controlling and rolling back an epidemic requires a carefully orchestrated, systematic, prioritized, multipronged effort to address each of its constituent elements. However, it is also recognized that significant progress or major advances can sometimes be precipitated by relatively minor interventions. Again, there are lessons and insights to be learned here for orchestrating a global counterterrorism campaign.

Ideology can be considered a type of meme or suite of memes. Memetics, the study of the contagiousness of thought and the dynamics that govern the spread of ideas, has captivated a broad audience for many years (e.g., Dawkins 1976; Lynch 1996; Gladwell 2000; Distin 2005). Dawkins (1976) defines memes as any cultural entity (e.g., a fashion fad, song, idea, religion, language) that is replicated through exposure to humans and has evolved as an efficient (though not necessarily perfect) copier of information and/or behavior. Much of the inquiry into memetics has centered on why some memes spread in an "epidemic" fashion. Like genes, memes spread if they provide a clear benefit to an individual, for example, through group identity, enhanced sexual attractiveness, or increased resource acquisition (Sober and Wilson 1998). Moreover, memes with no apparent benefit may arise and become fixed in a population simply because they possess characteristics that render them more likely to be adopted than another meme (e.g., incorrectly transmitted verbal phrases that contain more audible syllables than their grammatically correct counterparts [Dawkins 1989]). Even memes with a seemingly negative impact may spread through their provision of a net indirect benefit. For example, individuals with a "handicap" may be attractive to others due to the perception that the handicapped individual tolerates adverse conditions (Zahavi 1975) or actions that lead to the perception of a future benefit (rewards in the afterlife) can spread even if they do not increase fitness (see also Sosis and Alcorta, this volume). Here, we focus on terrorist ideology, but the analogy could apply to ideology in general. Society has often attempted to suppress memes such as political ideology, religion, and language, and the infectious disease analogy could serve to better understand these actions. However, while the infectious disease analogy could apply to ideology in general, our interest is specifically focused on the present broad-scale efforts to suppress terrorism ideology.

## The Infectiousness of Terrorist Ideology: Constructing the Analogy

The analogy takes shape at the system level, in the complex where infectious agents and terrorist ideologies exist (Fig. 12.1). Both entities require a unique set of conditions to remain established in a given system. Infectious agents depend on one or multiple hosts to support and complete their lifecycle (development, maturation, and reproduction). Host-pathogen dynamics can be viewed at the level of a host individual, host population, or community of species. Analogous to this, a given ideology can be "hosted" by a terrorist, terrorist cell, or terrorist organization. It is within these individuals and groups that ideology is conceived, formed, developed, and honed. Infectious agents need hosts just as ideology cannot exist without the minds that harbor it. In addition, pathogens and ideologies can persist in alternate forms outside of a host; many infectious agents have free-living resting stages and ideologies can be preserved outside the mind by various media.

Neither an infectious agent nor its host can exist in a system where external conditions are unsuitable. Indeed, abiotic (i.e., climatic) and biotic (i.e., competition, predation) factors help to shape the boundaries of an infectious organism's geographic range. The analog to this is the political,

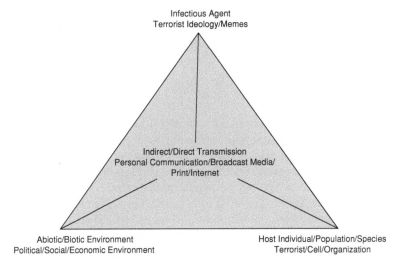

FIGURE 12.1 The analogy between the spread and dynamics of infectious agents and terrorist ideology is best depicted through a conceptual model of the systems where each entity exists. Adapted from Stares and Yacoubian (2005).

social, and economic environment that shapes the region where terrorist ideology evolves and thrives. Changes to the environment might foster the evolution of ideology away from that which promotes terrorism. For instance, the transition of a terrorist group to a recognized political power may lead to ideological or methodological changes better suited for formal governing (Prusher 2006).

Finally, it is the transmission of infectious agents and terrorist ideology between hosts/terrorists that maintains the entity's existence within the system. Infectious agents can be transmitted between susceptible hosts directly, as in the case of sexually transmitted pathogens, which require person-to-person contact. Transmission may also occur indirectly, by way of a vector that harbors the infectious agent (e.g., mosquitoes carrying malaria or ticks carrying Lyme disease), or through a contaminated vehicle such as food (e.g., salmonella) or liquid (e.g., giardiasis). The spread of terrorist ideology among individuals occurs via similar routes. Ideology may be transmitted directly, through the oral exchange of ideas from a terrorist to an individual who does not yet harbor the ideology. It may also spread indirectly, via a vehicle/vector such as broadcast media, print media, or the Internet.

Conceptual models, such as the one depicted in Fig. 12.1, are a common first step in understanding the framework of a system where an infectious agent exists. As we have shown here, the model is also well suited for deconstructing the complex where terrorist ideology thrives. Following the path of an investigative epidemiologist, our next task is to consider the quantitative models that best describe the spread of infectious agents/ terrorist ideology.

## Ecology and Epidemiology

Ecology and epidemiology have a highly developed set of theoretical tools and mathematical modeling approaches for understanding the basic properties of infectious agents. In many cases, such techniques have greatly informed control practices (Smith et al. 2005). Here we compile the theoretical framework from ecology and epidemiology that is commonly used as the basis for studying the spread of infectious agents. We extend these concepts to account for the establishment and spread of terrorist ideology and consider their utility in counterterrorism strategies. Johnson and J. Madin (this volume) similarly explore the relevance and applicability of models developed for fisheries stock assessment to the control of insurgent forces in unstable areas.

### Models

The most basic models of pathogen spread divide a population into two categories that reflect infection status: individuals susceptible (S) to or infected (I) with the pathogen (Fig. 12.2). SI models apply specifically to a group of infectious agents called microparasites, which include many viruses, bacteria, and protozoa. The rate of change in the abundance of individuals in each category (S or I) is described by a differential equation constructed from variables such as contact rate, susceptibility, birth rate, mortality, and the abundance of individuals in the host population. The frequency of a particular category can increase, decrease, or stabilize over time. A key to such models is that the differential equations are coupled. For simple formulations, it is possible to solve these equations analytically and gain general insight. In most cases, however, finding analytical solutions for coupled differential equations with more than two categories is mathematically intractable. In these cases, epidemiologists have turned to a very effective shortcut called the basic reproductive ratio, or $R_0$.

$R_0$ is the estimated number of secondary cases directly arising from one primary case (Anderson and May 1979). In deterministic models, if $R_0 < 1$, the epidemic will fizzle, while for $R_0 > 1$, an epidemic will occur. Once an epidemic begins, it grows at a rate known as the effective reproductive ratio, $R$, which tends to decline from $R_0$. $R$ can stabilize around 1 (become endemic in the population), or it can tend toward 0, at which point the pathogen is extirpated from the population (cycling, chaos, and other

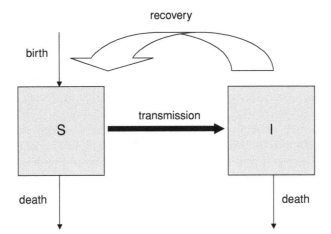

FIGURE 12.2 Flow diagram representing the basic SI (susceptible, infected) model used to study the spread of infectious agents. The model may be applied to the spread of terrorist ideology. Hosts/individuals are classified into one of the two categories. Young born into the susceptible category may come into contact with an infected individual/terrorist, transfer into this category, and either die from the infection/terrorist ideology or recover and return to the susceptible category.

complex outcomes are also possible). Determining the structure of the $R_0$ equation helps identify the variables (and their interactions) that are most likely to determine if infection will spread through an uninfected population.

Although an SI model tracks two classes of individuals over time, only the rate of change of the infected individuals is needed to calculate $R_0$. A simple component of an SI model that tracks the per capita rate of change of infected individuals (I) in an SI population can be written as

$$dI/Idt = S\beta - (\mu + \alpha + r),$$

where $S$ is the number of susceptible individuals, $\beta$ is the transmission coefficient (how contagious the infectious agent is to susceptibles), $\mu$ is the background mortality rate, $\alpha$ is the additional mortality rate suffered by infected individuals, and $r$ is the recovery rate. This model assumes that contacts occur at random and are density dependent (e.g., an individual living in a large city will come in contact with more people per day than an individual living in a rural community). $R_0$ is simply a ratio comprising the expected rate of transmission through contact with susceptible individuals divided by the sum of the death and recovery rates. The basic reproductive ratio is, therefore,

$$R_0 = S\beta/(\mu + \alpha + r).$$

By analogy, individuals that are exposed to and "infected" with terrorist ideology become terrorists. Some part of the remaining individuals in the population, the nonterrorists, are susceptible to the ideology. Understanding the components of $R_0$ for the spread of terrorist ideology may provide greater insight into the types of counterterror strategies that could be used to reduce $R_0$ below 1 and thus diminish the likelihood that terrorist ideology will spread. Efforts to reduce $R_0$ below 1 may concentrate on increasing the death rate of "infecteds," decreasing the abundance (or susceptibility) of susceptibles, or decreasing contact rate between the two. This seems, at first glance, to be a very clean analogy with obvious application. There is, however, an important caveat: the type of pathogen that terrorist ideology is compared to may fundamentally alter the predictions of the analogy.

## Terrorist Ideology: A Virus or a Worm?

Parasitologists realized that parasitic worms were not well modeled by the SI equations developed for viruses. This is because hosts exposed to many worms suffer higher pathology than hosts exposed to few worms. Hosts with many worms also contribute more to the growth of the total worm population. In addition, few infected hosts have the average level of infection: most have light, nonpathogenic infections, and the majority of the worms are in a few highly infected individuals (such distribution of parasites in a host population is referred to as aggregation). For this reason, the SI categorization of hosts as infected and uninfected fails to capture important aspects of the population dynamics of parasitic worms. New "macroparasite" (worms) models were developed to account for this level of biological detail (Crofton 1971; Anderson and May 1979; May and Anderson 1979). These models provided considerable insight into how to target control efforts toward the few, most heavily infected individuals.

As with parasitic worms, the spread of terrorist ideology may not be well explained by SI models. In particular, if ideological fervor can be considered a continuous variable, then some segments of a population may possess none of the ideology, some may posses a little, and a few may subscribe with high intensity. Individuals that subscribe most to the ideology may be more likely to spread the ideology than do those that are merely sympathetic or supportive of the ideology. This opens the question as to whether variation in the intensity of terrorist ideology among individuals is sufficiently important to justify models that are more complicated. If so, it is important to note that macroparasite models only indirectly account for the intensity of infection. Such models do not distinguish infected from uninfected hosts. Instead of tracking infected hosts, they track the total worm population. In applying macroparasite models to terrorism, the currency tracked would be the amount of terrorist ideology in the population, not the number of individuals subscribing to the terrorist ideology. The population level of terrorist ideology would have to be measured by various proxies (e.g., attack rate or internet chatter). The differential equation representing the growth of the worm population has a number of new terms. In a very simple formulation (ignoring here the change in the host population), the per capita rate of change of the adult worm population is

$$dW/Wdt = IS\beta/W - (\mu + \alpha + d + \alpha W(k + 1)/Sk),$$

while the per capita rate of change in infectious stages is

$$dI/Idt = \gamma W/I - \phi - \beta S,$$

where the variables are the same as above except that $W$ is the total number of worms in the worm population, $I$ is the total number of infectious stages, $\beta$ represents contact between infective stages and hosts, $d$ is the death rate of adult worms within the host, $\gamma$ is the rate that worms produce infective stages, $\phi$ is the death rate of infective stags in the environment, and $k$ is an inverse measure of the degree of aggregation of worms in the host population. Solving for the basic reproductive ratio of a macroparasite is much more difficult than for a microparasite, and alternative formulations for $R_0$ in macroparasites can be considerably different, depending on the biological details of the model (Roberts et al. 2002). One approach is that $R_0$ is the product of the mean number of new infections produced by a single adult parasite and the average life expectancy of adult and larval stages, or

$$R_0 = S\beta\gamma/[(\mu + \alpha + d)(\phi + S\beta)].$$

Clearly, $R_0$ increases with the lifespan of adult worms. Therefore, one can use chemotherapy to reduce the spread of disease. The spread of ideology could similarly be controlled by changing the ideology of terrorists, or, as seen in the microparasite models, reducing terrorist life span. An additional insight from this formulation not found in microparasite models is that $R_0$ is sensitive to the lifespan and production of infective stages. This is not often possible to control in parasite control programs, but the ideological analogue of infective stages (writings, media, and direct communication) may be possible to target.

An alternative to a macroparasite model is an SI model where contact rate varies among individuals. This has been of particular interest in understanding the spread of the HIV virus because the number of sexual partners varies greatly among individuals, with most of the HIV transmission stemming from a small number of highly sexually active people (Johnson et al. 1989). Similarly, some individuals who subscribe to terrorist ideology may actively make contact with others to promote its spread, while others are more private in their beliefs and thus contribute less to transmission. The derivation of models applied to this scenario is beyond the scope of this chapter, but the important result is that variation in contact rate among individuals can greatly increase $R_0$ (May and Anderson 1988). This suggests that targeting individuals who are disproportionately active will reduce the spread of terrorist ideology, a conclusion that seems intuitive. However, these models describe differences in contact rate that are distinctive to an individual whether or not he or she is infected (i.e., the link between infection and the propensity to spread an infection is indirect), which is different from the macroparasite models, which stipulate a direct link between infection intensity and spread of infection. It is not necessarily clear which approach provides a better analogy for terrorism. We attempt to tackle this conundrum by deconstructing and examining, in detail, the major contributing variables in the models: susceptibility, contact rate, recovery, and mortality.

## Extending Epidemiological Variables to the Spread of Terrorist Ideology

### Susceptibility

Without susceptibles, an infectious agent cannot spread. Although some infectious agents are able to evolve adaptations to get past host defenses, they are less likely to adapt to two host species that differ greatly in physiology and/or evolutionary history (Combes 2001). As a result, not all host species are susceptible to all pathogens, and host specificity, or the restriction of the number of host species that a pathogen can infect, is widespread.

Ideological specificity might arise for the same reasons as host specificity. The specific cultural identity of individuals (religious, national, historical) may predispose them to some, but not other ideologies. If an ideology is recognized as clearly foreign, an individual may more likely reject it as "nonself." Xenophobia is a common aspect of human cultures that probably acts to maintain cultural identity and prevent the spread of foreign ideologies (Boyd and Richerson 2005). For this reason, an ideology that is successful in spreading through a particular population may do so because it contains attributes compatible with that population's culture. In contrast, a population may be less susceptible to a terrorist ideology if that ideology differs greatly from the culture's modal ideology. This suggests that the cross-cultural spread of terrorist ideology may be unusual, and it is probably not reasonable to assume that all cultures are equally susceptible to a particular ideology—just as host species are differentially susceptible to infectious agents. While some terrorist ideologies span cultures, others take vastly different forms between cultures, or the same goals are rationalized in different ways salient to each culture.

A second pattern in susceptibility to infectious agents is that for a single species (or closely related species), populations that have had an evolutionary history often evolve adaptations against an infectious agent, while naïve populations can be more susceptible and less tolerant. The pressures that result in the evolution of resistance derive from the negative fitness consequences of an infectious agent on its host. This suggests that nations with a history of battling terrorism might be more likely to resist the future spread of terrorist ideology within their population.

The environment can also alter a host's susceptibility to an infectious agent. Thermal stress, for example, appears to increase the susceptibility of some marine organisms to different infectious agents (Lafferty et al. 2004). Other physiological stressors, such as lack of food, may force a host to redistribute resources away from defense against infectious agents (Rigby and Moret 2000). While stress at the individual level should tend to increase susceptibility to infectious agents, it may, unexpectedly, reduce the spread of a pathogen through a population. This occurs for the following reasons: stress may reduce host abundance (and, therefore, contact rates), and stress can increase pathogen mortality rates either through killing pathogens directly or by leading to differentially high mortality of infected hosts (Lafferty and Holt 2003). The environment could similarly affect the number of individuals susceptible to terrorist ideology if personal stress is a result (unemployment, reduced mating opportunities, frustration, humiliation, revenge). In environments that are unfavorable (poor in justice, resources, or freedom), new terrorist ideologies might hold appeal, particularly if they offer the promise of change, and current or alternative ideologies do not appear effective.

## Contact Rate

Pathogens and ideologies spread primarily through contact. SI models assume that in dense populations, individuals are more frequently in physical contact with one another. Another aspect of contact is the extent of movement among populations. This is difficult to capture in simple epidemiological models, but the logic is straightforward: contact between individuals of different populations is likely to lead to a wider spatial spread of an infectious agent. This is why increased contact through diffuse networks related to increases in transportation and modern trade has led to increasing concern for the spread of pandemics (Hufnagel et al. 2004). Current concern about avian influenza is one such example, and rightly so, as the

1918 avian flu pandemic may have been associated with large-scale troop movements during World War I (Barry 2004).

How similar is the spread of an ideology to the spread of an infectious agent? Whereas sick people do not consciously try to contact and infect other individuals (with the exception of some infectious diseases such as rabies), it is human nature to share ideas and convince others to agree with opinions. This is not to suggest that pathogen transmission is passive in comparison to ideological transmission. Infectious agents that are most likely to persist are those with traits enabling them to spread from one individual to another. Symptoms such as coughing, sneezing, and diarrhea are examples of behaviors induced by infectious agents to facilitate spread (Ewald 1993). It is also likely that ideologies are under analogous forms of selection for characteristics that favor spread (Dennett 2006). Successful religions often have doctrines favoring spread, such as active conversion of others, increased reproduction, early indoctrination, and retention (e.g., symbiont theories of the spread of religious ideologies). It has been suggested that such "symbiotic" relationships between religion memes (i.e., ideology) and their "hosts" may take the form of mutualism, in which both the meme itself and the host benefit, or they may be parasitic, in which the host is in some way oppressed by the religious ideology while the meme itself benefits (Dennett 2006). We may expect successful terrorist ideologies to possess adaptive traits for spread or to be aligned with existing religious ideologies. For example, terrorist ideologues have been successful at recruiting members through religious schools (Gunaratna 2005).

There are different opportunities for the spread of an ideology than for the spread of an infectious disease. Electronic communication greatly facilitates the spread of information at speeds and scales far exceeding physical contact. Information contact is increasing due to fewer language barriers, landline and cellular phones, television, radio, and the Internet. This decouples the spread of an ideology from local population density (Ariza 2005) while simultaneously favoring decentralization and spatial spread. Terrorist cells, in particular, tend to communicate and exchange ideology and information primarily through the Internet. This is not to downplay the importance of direct communication. A study of enlistees into a terror network found that communication of ideologies primarily occurs horizontally, through immediate and secondary friends (~80%). The remaining 20% occurs vertically, through kinship ties (Sageman 2004). Following enlistment, individuals often self-organize into isolated cells, the preferred size of which is eight members.

## Recovery

Infected hosts frequently recover, often with the assistance of an immune system. Recovered individuals may be permanently immune to subsequent

infection. Anyone old enough to reflect on his or her childhood appreci-
ates that ideologies change over time, and it seems safe to assume that "terrorist" is not necessarily a permanent ideological state. In borrowing from
the SI model framework, we might divide individuals into distinct ideological classes of terrorist (infected) and nonterrorist (susceptible). While
conversion from nonterrorist to terrorist seems to be analogous to an
infectious process, recovery from terrorist to nonterrorist is not analogous
to an immune defense. In addition, while an individual's ideological intensity may wane or increase over time, a person does not necessarily adopt an
ideology then lose it over time to a nonideological state. Ideologies can
persist over time and may coexist with or interact with other ideologies. In
other words, there may be a potential series of "infected" states. This is particularly important for the concept of recovery because a change in ideological states would depend on the extent that individuals can be exposed
to alternative ideologies. In a sense, this is more like pitting two infectious
agents against one another than the effect of an immune system.

Although replacement is not the rule for infectious agents, we can turn
to flukes (parasitic trematode worms) for insight. Trematodes have complex life cycles that typically involve a snail as a first-intermediate host.
Once infected, snails rarely revert to an uninfected state. In many cases,
several species of trematode infect the same species of snail, but there is
generally only room for one trematode species at a time within an individual snail (few trematode species pairs can coexist). For this reason,
trematodes have developed special adaptations for battling with other
trematodes inside the snail, and, in many cases, one can predict which
trematode species will win the internal battle for host occupation (Kuris
and Lafferty 1994). If ideologies, like trematodes, have dominance hierarchies, facilitation of dominant ideologies could lead to permanent
"recovery."

## Mortality

All else being equal, highly pathogenic infectious agents have a smaller $R_0$
because infected individuals with short lifespans have fewer opportunities
to transmit an infection. A terrorist's actions may create an analogous situation. Even in the absence of counterterrorist efforts, terrorist mortality
rate should increase because militant terrorist activities are inherently dangerous (e.g., weapons training, handling explosives, suicide missions,
primitive living). These dangers may therefore make it more difficult for
terrorist ideology to spread. However, risky behavior and high mortality
rates may also increase ideological spread. This would be particularly true
if suicide attacks lead to martyrdom, benefits for the attacker's family, or a
desirable afterlife, or if mortality achieves a terrorist goal and inspires others to join.

Ecological and epidemiological models also suggest that removing infected individuals from the population can reduce the impact of the infectious agent. Although society does not permit the culling of infected patients, culling infected animals has been shown to be effective in preventing the spread of veterinary and zoonotic pathogens (recent examples include foot and mouth, avian influenza, and Nipah virus) (Barlow 1996). Analogously, counterterrorism efforts often attempt to remove terrorists. This may be motivated by a sense of justice, to directly reduce threat, and/or to help reduce the spread of terrorist ideology. However, as implied previously, the mode of removing terrorists from the population might ironically enhance the spread of ideology through martyrdom and other factors. For instance, the killing of a terrorist might inspire susceptibles to adopt the very ideology counterterrorist operations aim to defuse. An SI equation that links death and transmission through martyrdom is

$$dI/I\,dt = S\beta(1 + \alpha m) - (\mu + \alpha + r),$$

and

$$R_0 = S\beta(1 + \alpha m)/(\mu + \alpha + r),$$

where the variables are as before, and $m$ (martyrdom) is the extent that the death of a terrorist (resulting from terrorist activity) increases transmission to susceptibles.

We might then wish to explore how lethal removal of terrorists affects the spread of terrorism. The partial derivative of $R_0$, with respect to $\alpha$, indicates the slope of the relationship between removal and spread. This will be positive for

$$m > 1/(r + \mu).$$

This means that where $m$ (martyrdom) is near zero, increasing terrorist death rate (such as through military intervention or suicide attacks) will always slow the spread of terrorist ideology. Nevertheless, with increasing $m$, $r$, or $\mu$, an increase in terrorist death rate can increase the spread of terrorism. This makes obvious sense for $m$. Less intuitive is how a high background mortality or recovery rate increases the value of martyrdom. This occurs because when the lifespan of a terrorist (physical or ideological) is sufficiently short, that person's premature loss from the terrorist population has little cost to the spread of the ideology. Therefore, understanding the strength of martyrdom relative to recovery rates and background survivorship might allow better evaluation of counterterrorist strategies.

## Control and Prevention

We have taken the first steps to determine the utility of basic ecological and epidemiological theory in understanding the spread of terrorist ideology. If terrorist ideology is analogous to an infectious agent, what insight can public health offer counterterrorism?

### Preventative Control

Outbreaks of infectious agents are oftentimes apparent in the early stages, when very few individuals are infected. This is a direct result of the incentive that an infected individual receives for reporting their condition and thus receiving treatment. This, in turn, increases the likelihood that public health officials will learn of enough independent infectious cases to implement control initiatives before the rate of spread increases. Not surprisingly, this scenario is in stark contrast to the spread of terrorist ideology. Indeed, the very practice of terrorism and the ideologies upon which it is based demands a high level of covertness outside of the immediate cell. Relative to pathogen outbreaks, such behavior makes it exceedingly difficult to forecast and control emerging terrorist ideologies, though this is not to suggest a complete lack of indicators. Symptoms of an emerging terrorist ideology may include an increase in attacks, the death or capture of militants, or the dissemination of information (Stares and Yacoubian 2005). It is important to recognize, however, that the inception of infectiousness is not necessarily correlated with the appearance of symptoms. This level of unpredictability is why public health favors prevention, particularly vaccination, as the primary control strategy.

Vaccination works by exposing uninfected individuals to dead or attenuated infectious agents, after which the immune system goes through the process of building specific antibodies that then lead to immunity. Even vaccinating only a fraction of susceptibles in the population can reduce $R_0$ below 1 (a phenomenon called herd immunity). Vaccination is most viable as a broad-scale control strategy when used against well-known pathogens. An analogy to vaccination for terrorist ideology is exposing a population to a discredited version of the ideology to make it less likely that newly exposed individuals will find it attractive (Stares and Yacoubian 2005). An in-depth knowledge of a particular terrorist ideology would seem necessary for this approach to succeed.

Although vaccination is available for many of the pathogens that plague mankind, the evolutionary novelty and unpredictability of emerging infectious agents dampens its viability as a broad-scale control strategy. In the absence of vaccines for pathogens on the verge of an outbreak, public health officials generally favor two basic control practices: (1) isolation of symptomatic individuals and (2) tracing and quarantining their contacts (Fraser et al. 2004). Both the implementation and success of these measures rely on a number of factors, ranging from the epidemiological characteristics of the infectious agent to the communication infrastructure of the public health agencies charged with control. Mathematical modeling of contemporary outbreaks (in particular SARS, HIV, smallpox, and influenza) suggests that the success of these control measures is equally dependent on the proportion of transmission that occurs prior to the onset of clinical symptoms ($\theta$) and the inherent transmissibility of the infectious agent ($R_0$ as described in the previous section of this chapter) (see Fraser et al. 2004 for model details). Model simulations suggest that isolation, contact tracing, and quarantine are indeed sufficient to control outbreaks of infections when the values of $\theta$ and $R_0$ fall below a critical threshold. Control through isolation alone is possible when $\theta < 1/R_0$, but when $\theta > 1/R_0$, contact tracing is also required. When $\theta$ is very high, neither control measure (alone or in combination) is able to prevent the outbreak from progressing.

Similar to emerging infectious diseases, the early detection of new terrorist ideologies may aid control. Due to the covert nature of terrorism, early detection is also extremely difficult. In the first three years following 9/11, the U.S. counterterrorism strategy was to target operational terrorist cells, as opposed to their ideological motivations. Coined "operational counterterrorism," the campaign monitored suspected terrorists, collaborators, supporters, and sympathizers and used the acquired intelligence to facilitate arrests (Gunaratna 2005). By targeting terrorists in the early stages of planning, operational counterterrorism reduced what may have resulted in longer-term terrorist activity (Gunaratna 2005). However, models of infectious pathogen outbreaks suggest that the success of counterterrorism will increase if detection occurs before the onset of the obvious indicators (attacks, media dissemination, etc.). As suggested by the pathogen outbreak model described above, if $\theta$ gets too high, even the most robust public health control measures become ineffective. There is reason to believe that the same would hold true for emerging terrorist ideologies.

Strategic counterterrorism seeks to reduce the political and operational space for terrorism to develop, spread, and sustain (Gunaratna 2005). The targets of strategic counterterrorism include the foundations on which a culture's education, religion, media, legislation, and ideology are set. The mission is to counter terrorism in these very institutions by promoting an ethic against terrorist ideology that extends to the greater community. Because strategic counterterrorism is in many ways preemptive, susceptible individuals are likely to become infected with an antiterrorism agenda before they are ever exposed to a terrorist ideology.

As discussed earlier, unfavorable environments (in terms of justice, resources, or freedom) may promote the adoption of new terrorist ideologies by those suffering personal stress. Therefore, efforts that reduce the desire for change or provide alternative pathways for change might decrease susceptibility. Still, environmental conditions that decrease the susceptibility of individuals to a particular ideology will not necessarily prevent the spread of that ideology. For example, improved economic conditions might decrease desire for change in individuals, but also increase opportunities for communication or implementation of a terrorist ideology, thus making it difficult to predict the net effect.

The study of pathogen outbreaks and emerging infectious agents has grown in recent years. In the midst of fear factors such as avian influenza H5N1, we can expect the trend to continue. To date, however, scientists have found it extremely difficult to identify the forces that would allow public health officials to forecast the size or timing of pathogen outbreaks. Not surprisingly, this is because the ultimate dynamics of pathogen outbreaks are highly influenced by numerous factors that are in no way static: the social structure and immunology of the population, the epidemiological characteristics of the infectious agent, the environment, and contact rates between infecteds and susceptibles. Despite these difficulties, scientists are getting better at accounting for these factors and thus reducing the forecast envelope for the size and timing of pathogen outbreaks (Fraser et al. 2004; Hufnagel et al. 2004; Drake 2006). As progression in this field continues, we should be able to add even more dimensionality and depth to the infectious agent–terrorist ideology analogy.

## Postestablishment Control

Thus far we have discussed the value of prevention as a tool to control the emergence and spread of infectious agents and, analogously, terrorist ideology. How do public health officials attack infectious agents that are already established, and are there counterterrorism analogues?

Efforts to reduce contact can be effective at slowing the spread of infectious agents when they have become established. Two general approaches are used. The first is to reduce contact rates between individuals irrespective of their infection status. Safe-sex and personal-hygiene campaigns are probably the best examples in modern society. For instance, children are taught to wash their hands, use toilets, and avoid spitting in public. The second approach is to specifically reduce contact between infected and uninfected individuals. In public schools, children with head lice are typically sent home to reduce the spread of lice to their classmates. For more serious infectious diseases, public health officials may quarantine infected individuals (and sometimes those that the infected individuals have contacted) in an effort to stop spread (Fraser et al. 2004). Such an approach was used to reduce the spread of SARS. Limits on the movement of infected individuals can be difficult because detection is challenging. In the SARS epidemic, body temperature scans in airports (e.g., Hong Kong) were used to identify individuals with fevers, and these individuals were then subjected to medical examination. Recent epidemiological models are making progress in predicting the spread of pathogens such as SARS with evaluation of different control strategies (Hufnagel et al. 2004).

As mentioned earlier, controlling the spread of an infectious disease can occur through reducing contacts in general or isolating infected individuals. Limiting communication within a population might reduce the spread of a terrorist ideology but could also limit the spread of counterterrorist ideology, as well as increase resentment by denying expected freedoms of expression. Identifying and isolating individuals "infected" with terrorist ideology may be a more practical means of reducing spread. While covertness greatly impairs the ability to forecast and control emerging terrorist ideologies, detectable symptoms of an emerging terrorist ideology may include an increase in attacks, the death or capture of militants, or the dissemination of information (Stares and Yacoubian 2005). Active individuals might be identified via their own efforts to communicate their ideology or through identification of existing networks. Strategies presently used to reduce the ability of ideologues to contact others include imprisonment, military isolation, and disruption of communication, among others.

For some infectious agents (e.g., ebola, west Nile, rabies), contact with a reservoir host species drives transmission dynamics. Controlling such pathogens in peripheral hosts, such as humans, is greatly hindered because the pathogen is primarily supported in nonhuman hosts, which may be outside the reach of control efforts. Moreover, when the reservoir host suffers little pathology, there is no selection on the infectious agent to evolve reduced virulence. This is the case for many of the zoonotic infectious agents that spill over from wildlife reservoirs and cause high rates of morbidity and mortality in susceptible human populations.

An analog exists in the spread of terrorist ideology. Individuals that abstain from the practice of terrorism may still sustain the founding ideology, which they actively fund, harbor, and spread to recruit active terrorists. In doing so, they become reservoirs for the ideology. Like peripheral hosts, the individuals they recruit are more likely to exhibit the symptoms of the ideology (terrorist acts) than they are to spread the ideology or participate in recruitment activities. Decoupling terrorist recruitment from terrorist action makes it less likely that operational counterterrorism specifically directed against terrorists and their activities will reduce the spread of terrorism. In this scenario, the reservoir individuals harboring the ideology would be undertargeted. Terrorist networks with an important reservoir might be better modeled as an infectious pathogen with a reservoir host, with counterterrorism actions targeted specifically against the reservoir.

## Discussion

With this chapter, we have presented a framework and set of models as alternative means of conceptualizing the nature of terrorism. The epidemiological approach and tools presented here form the basis for understanding, controlling, and predicting the spread of infectious pathogens in human and wildlife populations. By drawing parallels between the spread of infectious agents and the spread of terrorist ideology, we have sought to highlight both the similarities and contrasts between these two threats. In so doing, we suggest how the vast body of knowledge in ecology and epidemiology may be modified and potentially applied to understanding terrorism.

In some cases, basic strategies for controlling infectious agents may translate directly to controlling terrorism. In particular, ideological analogs to $R_0$ could be used with coupled differential equations to predict the spread of terrorist ideology. As we have shown, the success of this approach is strongly dependent upon the variables used to construct the equations. Existing generic epidemiological models are not adequately suited for modeling terrorist ideology, just as they are not often adequately suited for modeling specific pathogens. Indeed, a good deal of critical thought will be required to construct adequate definitions for individuals that are susceptible to and infected with terrorist ideology, and for the model parameters that link these stages (e.g., transmission rate, death rate, and aggregation of the ideology, as well as contact rate between individuals). These variable definitions and associated assumptions would ideally be sufficiently analogous to those in the epidemiological models on which they are based without losing the social dimensions that make them inherently unique.

Simple epidemiological models could be customized for terrorist ideology by incorporating a sufficient amount of detail. Such model adaptation might consider the following factors. The types of infected individuals (terrorists) might be broken down into additional categories lacking analogues in epidemiological models (e.g., militants, educators, leaders). One might consider the possibility of vertical (parent-offspring) transmission of terrorist ideology. Unlike infectious agents that attack a host and, in turn, are attacked by the host's immune system, infectious ideologies may invade a host that already harbors an ideology. An existing ideology might resist invasion or it may be replaced or altered. It seems that models of terrorist ideology would benefit from consideration of heterogeneities in human behavior relating to spread. It would be worth considering how to incorporate terrorist intensity but, like macroparasite models, identify a currency (e.g., attack rate) useful for counterterrorist goals. Once adequate models were identified, it would be possible to determine what sorts of data might be suitable for better tracking the spread of terrorist ideology. Moreover, there is much to be gained from the study of novel pathogen outbreaks and the models used to predict their timing, size, and geography. Collaboration between ecologists, epidemiologists, and social scientists has proven beneficial to the study of various human pathogens. Keeping with the analogy, such forms of cross-disciplinary collaboration might also benefit the study of terrorism.

## Summary

Goals for future research include the following:

Develop working definitions and assumptions for the variables and parameters that characterize SIR models of terrorist ideology.

Determine how behavioral, sociological, and cultural parameters would be accounted for in these models.

Consider how the strength of these models vary when temporal (e.g., political cycles) and spatial (e.g., global terrorist networks vs. local terrorist cells) scales are accounted for.

Test the models against analytical data on the spread of ideologies.

The following questions remain open:

How will ecological and epidemiological models of infectious agent spread be modified to accommodate the diversity of terrorist ideologies?

What are the ethical considerations of comparing terrorist ideology to an infectious agent and, further, using this analogy to construct counterterror campaigns?

Are the strengths and benefits of the infectious agent–terrorist ideology analogy sufficient to warrant continued pursuit of its utility?

How might operational and strategic counterterrorism use insights from the infectious agent–terrorist ideology analogy?

ACKNOWLEDGMENTS

This work was conducted as part of the Ecological and Evolutionary Models for Homeland Security Strategy Working Group supported by the National Center for Ecological Analysis and Synthesis, the University of California, and the University of California, Santa Barbara. This manuscript has benefited from support received from the National Science Foundation through the NIH/NSF Ecology of Infectious Disease Program (DEB-0224565). Dominic Johnson, Bradley Thayer, and Geerat Vermeij provided useful comments on a draft of this chapter.

## REFERENCES

Anderson, R. M., and R. M. May. 1979. Population biology of infectious diseases. Part 1. *Nature* 280: 361–367.

Ariza, L. M. 2006. The Internet as the ideal terrorism tool. *Scientific American* January, 2006. http://www.sciam.com/article.cfm?chanID=sa006&co/ID=17&articleID=000B5155-2077-13A8-9E4D8341B7F0101.

Barlow, N. D. 1996. The ecology of wildlife disease control: Simple models revisited. *Journal of Applied Ecology* 33: 303–314.

Barry, J. M. 2004. *The great influenza.* New York: Penguin.

Boyd, R., and P. J. Richerson. 2005. *The origin and evolution of cultures.* New York: Oxford University Press.

Combes, C. 2001. *Parasitism: The ecology and evolution of intimate interactions.* Chicago: University of Chicago Press.

Crofton, H. D. 1971. A model of host-parasite relationships. *Parasitology* 63: 343–364.

Dawkins, R. 1976. *The selfish gene.* Oxford: Oxford University Press.

Dawkins R. 1989. *The selfish gene,* 2nd ed. Oxford: Oxford University Press.

Dennett, D. C. 2006. *Breaking the spell: Religion as a natural phenomenon.* New York: Viking Penguin.

Distin, K. 2005. *The selfish neme: A critical reassessment.* Cambridge Press.

Draks, J. M. 2006. Limits to forecasting precision for outbreaks of directly transmitted diseases. *PLoS Medicine* 3: 57–62.

Ehrlich P. R., and S. A. Levia. 2005. The evolution of norms. *PLoS Biology.* 3:e194.

Ewald, P. W. 1993. *Evolution of infectious disease.* Oxford: Oxford Press.

Fraser, C., S. Riley, R. M. Anderson, and N. M. Ferguson. 2004. Factors that make an infectious disease outbreak controllable. *PNAS* 101:6146–6151.

Gladwell, M. 2000. *The tipping point: How little things can make a big difference.* Back Bay Books.

Gunaratna, R., ed. 2005. *The changing face of terrorism.* Singapore: Academic.

Hufnagel, L., D. Brockmann, and T. Geisel. 2004. Forecast and control of epidemics in a globalized world. *Proceedings of the National Academy of Sciences USA* 101: 15124–15129.

Johnson, A. M., J. Wadsworth, P. Elliot, L. Prior, P. Wallace, S. Blower, N. L. Webb, G. I. Heald, D. L. Miller, M. W. Adler, and R. M. Anderson. 1989. A pilot study of sexual lifestyle in a random sample of the population of Great Britain. *AIDS* 3: 34–142.

Kuris, A. M., and K. D. Lafferty. 1994. Community structure: Larval trematodes in snail hosts. *Annual Review of Ecology and Systematics* 25: 189–217.

Lafferty, K. D., and R. D. Holt. 2003. How should environmental stress affect the population dynamics of disease? *Ecology Letters* 6: 654–664.

Lafferty, K. D., J. Porter, and S. E. Ford. 2004. Are diseases increasing in the ocean? *Annual Review of Ecology, Evolution, and Systematics* 35: 31–54.

Lynch, A. 1996. *Thought contagion: How belief spreads through society. The new science of memes.* New York: Basic Books.

May, R. M., and R. M. Anderson. 1979. Population biology of infectious diseases. Part II. *Nature* 280: 455–461.

May, R. M., and R. M. Anderson. 1988. The transmission dynamics of human immunodeficiency virus (HIV). *Philosophical Transactions of the Royal Society B* 321: 565–607.

Office of Homeland Security. 2002 National Strategy for Homeland Security. Washington, D.C.: Office of the President of the United States of America.

Prusher, I. R. 2006. Will Hamas change course? *Christian Science Monitor*, February.

Rigby, M. C., and Y. Moret. 2000. Life-history trade-offs and immune defenses. In *Evolutionary biology of host-parasite relationships: Theory meets reality*, ed. R. Poulin, S. Morand, and A. Skorping, 129–142. Amsterdam: Elsevier.

Roberts, M. G., A. P. Dobson, P. Arneberg, G. A. de Leo, R. C. Krecek, M. T. Manfredi, P. Lanfranchi, and E. Zaffaroni. 2002. Parasite community ecology and biodiversity. In *The ecology of wildlife diseases*, ed. P. J. Hudson, A. Rizzoli, B. T. Grenfell, H. Heesterbeek, and A. P. Dobson. Oxford: Oxford Biology 63–82.

Sageman, M. 2004. *Understanding terror networks*. Philadelphia: University of Pennsylvania Press.

Smith, K. F., A. P. Dobson, F. McKenzie, L. Real, D. Smith, and M. Wilson. 2005. Ecological theory to enhance infectious disease control and public health policy. *Frontiers in Ecology and the Environment* 3: 29–37.

Sober, E., and D. S. Wilson. 1998. *Unto others*. Cambridge, MA: Harvard University Press.

Stares, P., and M. Yacoubian. 2005. Terrorism as virus. *Washington Post*, August 23.

Zahavi, A. 1975. Mate selection: A selection for a handicap. *Journal of Theoretical Biology* 53: 205–214.

Part Six

# SYNTHESIS

Chapter 13

# PARADIGM SHIFTS IN SECURITY STRATEGY

Why Does It Take Disasters to Trigger Change?

DOMINIC D. P. JOHNSON AND ELIZABETH M. P. MADIN

*If men could learn from history, what lessons it might teach us! But
passion and party blind our eyes, and the light which experience gives
is a lantern on the stern, which shines only on the waves behind us.*
SAMUEL COLERIDGE

Prior to 9/11, U.S. counterterrorism policy and intelligence suffered
from numerous problems. The striking feature about this is not the flaws
themselves, but rather that these flaws were long appreciated and nothing
was done to correct them. It took a massive disaster—3000 American
deaths—to cough up the cash and motivation to address what was already
by that time a longstanding threat of a major terrorist attack on the U.S.
homeland.

A second striking feature is that this failure to adapt is no novelty. Pearl
Harbor, the Cuban Missile Crisis, and the Vietnam War were also belated
wake up calls to adapt to what in each period had become major new challenges for the United States.

Just as the military is accused of "fighting the last war," nations fail to
adapt to novel security threats. The status quo persists until a significant
number of lives or dollars are lost. Only at these times can we be sure that
nations, institutions, and elected representatives will fully adapt to novel
security threats. If we understand why this is so, we will be better able to
avoid further disasters in the future.

We suggest that it takes disasters to trigger change because (1) dangers
that remain hypothetical fail to trigger appropriate sensory responses,
(2) psychological biases serve to maintain the status quo, (3) dominant
leaders entrench their own idiosyncratic policy preferences, (4) organizational behavior and bureaucratic processes resist change, and (5) electoral
politics offers little incentive for expensive and disruptive preparation for
unlikely and often invisible threats.

## The Curse of the Status Quo

Even a highly adaptable state might be able to prevent only 99 out of 100 disasters from happening. Such successes rarely make the news. By contrast, the 1% of disasters that do occur will be dramatic and visible and may therefore attract undue attention. Even so, the argument of this chapter is that human nature, and the nature of the institutions that humans create, exhibits a number of self-defeating phenomena that impede efficient adaptation to novel security threats, increasing the probability of periodic disasters. Indeed, under certain unfavorable conditions, we may be pathologically and institutionally unable to avoid disasters. This curse is sustained by a number of biases rooted in biology, psychology, advocacy, organizational behavior, and politics, all of which converge to preserve the status quo.

The same phenomenon is evident in everyday life. Accident-prone highways, dangerous machinery, or hazardous flight paths are often not altered until after significant numbers of people are killed or injured, or significant financial losses are incurred. In one sense this is logical: since disasters are hard to predict, only cumulative data exposes whether the costs of occasional disasters outweigh the costs of change (Perrow 1999). However, this logic is often flawed or inapplicable for two reasons. First, the costs of disasters, if they are measured in human lives, may be unacceptable. We cannot simply wait to see if or how often they happen. Second, the costs of disasters, however they are measured, are often known beforehand to outweigh the costs of not acting, yet still nothing is done to prevent them.

This phenomenon has parallels in other disciplines, including the history of science, epistemology, policy analysis, and economics, suggesting that it is a common denominator of human nature and human institutions, not something specific to a particular issue, culture, or context. For example, Thomas Kuhn described how scientific progress is characterized by lengthy periods of relative stasis where established models reign supreme, but that this status quo is punctuated by "paradigm shifts" that follow from exceptional findings such as those by Gallileo or Einstein (Kuhn 1970). Similarly, Michael Foucault argued that history itself does not proceed smoothly as a steady, linear continuum but is defined, rather, by moments of rupture that overturn prevailing systems of knowledge (Foucault 1970, 1977). Another example is the "punctuated equilibrium" theory in policy analysis, which describes how U.S. domestic policy follows periods of relative stasis during which decision processes and bureaucracies act to preserve a status quo, but this is punctuated by major periods of reform following the adoption of innovations, attention-riveting external events that grab government or public attention, and windows of opportunity when conducive factors coincide or when advocacy groups rise to prominence (Baumgartner and Jones 1993, 2002; Busenberg 2003). Finally, economics is famous for its quip that the field progresses only with each funeral.

Individuals corroborate, advertise, and propagate their favored theories as they grow older and more powerful. Only when they are gone can fresh alternatives take solid root. The same is true of other disciplines and organizations.

In many aspects of human endeavor, it appears that we fail to adapt to changing circumstances until there is a major event that wrenches us from established paradigms. We argue that this failure to adapt is, if anything, more likely in the domain of international politics than other domains because the ambiguity inherent in judgments of other cultures, ideologies, and motives allows false interpretations to prosper and persist at especially high levels (Johnson and Tierney 2006). We are, simply put, doomed to periodic foreign policy disasters.

The good news is that research in biology, psychology, organizational behavior, and political science reveal systematic causes of this phenomenon, offering the opportunity to predict when and where it will occur, and ways to correct it in the future. Policy makers may therein find ways to improve national security as well as maximize public and congressional support. Before expanding on the biases at work, we outline a series of events that illustrate the failure to adapt to novel security threats: the 1941 attack on Pearl Harbor, the 1962 Cuban Missile Crisis, the Vietnam War, and the terrorist attacks of 9/11.

## Examples of Disasters Triggering Change

The Japanese attack on Pearl Harbor in December 1941 is widely regarded as a colossal U.S. intelligence failure spanning the lowest to the highest levels of command (Iriye 1999; Kahn 1999). The striking thing is not that U.S. intelligence and strategic posture were inadequate, it is that they were known to be inadequate and yet failed to be changed. Although U.S. intelligence had no specific information or dates regarding the raid on Pearl Harbor, Japanese diplomatic codes had been broken, and a number of sources pointed to the likelihood of some kind of Japanese attack on the United States. It was the failure of the U.S. government to recognize the changing motives and intentions of the Japanese decision makers that led to a poor level of readiness in the U.S. Pacific Fleet. These inadequacies reflect a status quo bias in U.S. strategy toward Japan in the prewar period, summed up by historian of intelligence David Kahn (1999, 166):

> American officials did not think Japan would attack their country. To start war with so superior a power would be to commit national hara-kiri [suicide]. To Western modes of thought, it made no sense. This rationalism was paralleled by a racism that led Americans to underrate Japanese abilities and will. Such views were held not only by common bigots but by opinion-makers as well. These preconceptions blocked out of American minds the possibility that Japan would attack an American possession. . . . An attack on Pearl Harbor was seen as all but excluded. Though senior army and navy officers knew that

Japan had often started wars with surprise attacks, and though the naval air defense plan for Hawaii warned of a dawn assault, officials also knew that the base was the nation's best defended and that the fleet had been stationed that far west not to attract, but to deter, Japan.

Having committed errors of planning and intelligence that heightened both the probability and severity of the Pearl Harbor attack, the shock and moral outrage following the "day of infamy" led to major changes in U.S. security strategy. The entire foundations of U.S. intelligence were uprooted. The National Security Act of 1947 established the Department of Defense, the National Security Council, and the Central Intelligence Agency, in large part to ensure the integration of military and diplomatic intelligence so that such a disaster could never befall the country again. The Pearl Harbor disaster was exacerbated by the status quo bias in U.S. policy, but the shock of the attack itself caused a paradigm shift in U.S. security strategy.

The Cuban Missile Crisis of 1962 also represented a massive failure of U.S. intelligence (Allison and Zelikow 1999). When Soviet SS-4 and SS-5 missile sites were discovered on the island in October 1962, it sparked a major diplomatic crisis and military standoff in which the superpowers came perilously close to war. "American leaders," wrote Robert Jervis, "were taken by surprise in October 1962 because they thought it was clear to the Soviet Union that placing missiles in Cuba would not be tolerated" (Jervis 1983, 28). The U.S. deterrence strategy, in other words, had failed. War was in the end averted through a negotiated agreement, but the popular memory of U.S. victory masks the significant concessions that the United States also made, and the brinkmanship that could so easily have resulted in war (Johnson and Tierney 2004). Khrushchev is widely regarded, by Soviet as well as western contemporaries and historians, as having taken an enormous risk in deploying missiles on Cuba (Lebow 1981; Fursenko and Naftali 1997). In the face of such extreme risk taking, U.S. deterrence was based on faulty premises. The crisis sparked significant changes in U.S. policy, including opening direct lines of communication between the White House and the Kremlin, and a major restructuring of chain of command authority in the U.S. military (including the President's control over nuclear weapons). The Cuban Missile Crisis was exacerbated by the status quo bias in U.S. policy, but the shock of the crisis itself caused a paradigm shift in U.S. Cold War security strategy.

The Vietnam War also represented a failure of U.S. policy and intelligence. Policy suffered from the Cold War obsession with halting the spread of communism and failed to address the root cause of the insurgency as a war of national liberation (Gilbert 2002). Military strategy suffered because it sought to replicate traditional tactics of open combat. Intelligence suffered because it focused on conventional war metrics, such as body counts and weapons captured, and only belatedly shifted to address the key elements of nationalist sentiment and counterinsurgency (Gartner 1997). The realities of guerilla war were widely understood after the experience of the British in Malaya (1948–1960) and the French in Vietnam (1946–1954), but this had little impact on U.S. policy. The U.S. leaders believed that the gradual escalation of American military power combined with coercive diplomacy, which seemed to have worked well in the past, would work just as well in Vietnam. President Lyndon B. Johnson's press secretary, Bill Moyers, said after resigning in 1967 that in Johnson's inner circle "there was a confidence, it was never bragged about, it was just there—a residue, perhaps of the confrontation over the missiles in Cuba—that when the chips were really down, the other people would fold" (Janis 1972, 120). It came as a major shock for the United States to lose a war for the first time in its history. Following the withdrawal of U.S. troops in 1973, and the fall of Saigon in 1975, the "Vietnam syndrome" made the U.S. public, Congress, and subsequent administrations especially wary of military intervention overseas (limiting the country to small-scale actions, such as in Grenada and Panama). When the next big confrontation did occur, the 1991 Persian Gulf War, the Powell doctrine of overwhelming force and limited military objectives represented an enormous shift in strategy (Powell 1995). The Vietnam War was exacerbated by the status quo bias in U.S. policy, but the shock of defeat caused a paradigm shift in U.S. foreign policy with a legacy that survives to this day.

This now familiar pattern repeated itself on September 11, 2001. As William Rosenau put it, "although some policymakers and analysts have tried, it is impossible to deny that the events of 11 September 2001 represented a massive failure of intelligence" (Rosenau 2007, 143). The 9/11 commission and other sources reveal that a major terrorist attack on the U.S. homeland was by no means unexpected (Simon and Benjamin 2000; 9/11 Commission 2004; Clarke 2004). Intelligence agencies and counterterrorism experts had long argued that al-Qaeda presented a growing and significant threat in the 1990s—indeed, major terrorist plots of the scale of 9/11 had already been averted—but U.S. policy makers failed to adapt to meet this new threat (Gellman 2002; Rosenau 2007). In a replica of Pearl Harbor, the precise timing and method of attack was of course not predicted, but not preparing for an attack of this kind was the result of a huge intelligence failure. The structure and function of government agencies, as well as many key individuals, were stuck in a Cold War mindset, and had not adjusted adequately to the new threats of transnational terrorism. It took 9/11 to set in motion—too late of course—sweeping changes of government and intelligence organization that many had clamored for years to achieve (the U.S. Commission on National Security for the Twenty-first Century, for example, had warned of terrorist attacks on the United States in early 2001 and recommended the creation of a Department of Homeland Security). Today, "combating al-Qaida has become the central organizing principle of U.S. national security policy" (Rosenau 2007, 134). Why did it take 9/11 to get it there?

## Common Patterns

Although the examples above have much to distinguish them—different periods, locations, opponents, ideologies, geopolitics, and administrations—they nevertheless share common properties. In each event (1) the United States was faced with a novel threat, (2) the potential consequences of this threat were evident, and (3) the United States failed to adapt to this new threat. Nor are these cases anomalies in an ocean of otherwise efficient adaptation; numerous other such cases throughout history could fill several volumes (see, e.g., Dixon 1976; Perlmutter 1978; Snyder 1984; Tuchman 1984; Gabriel 1986; Cohen and Gooch 1991; Regan 1993; Perry 1996; David 1997; Hughes-Wilson 1999). All sides in World War I expected the war to be short and victorious, despite copious evidence to the contrary, and only the carnage of the war itself brought the end of an era in military thinking and the establishment of the League of Nations (Snyder 1984). In the 1930s, the allies thought Hitler had limited goals, despite his accumulating gains, and the horrors of World War II led to the dismemberment of Germany, an open-ended commitment to U.S. military deployments overseas, and the establishment of the United Nations. Similarly, the U.S. reliance on the use of force as a tool of policy was significantly curtailed by the Presidential War Powers Act (triggered by the shock of defeat in Vietnam), and the Goldwater-Nicholls Department of Defense Reorganization Act (triggered by the failed Iranian hostage rescue attempt in 1980). The need for these changes was well appreciated long before they came about, but only major disasters actually made them happen.

It is not only the United States that is subject to these failures. The same phenomenon is common in the history other nations. For example, the 1973 Yom Kippur War exposed a massive failure of Israeli intelligence. There were numerous warning signs of a joint Egyptian and Syrian attack that Israeli military and political leaders failed to acknowledge (Blum 2003; Rabinovich 2004). Following the hugely successful 1967 Six-Day War, and Israeli preconceptions of what it would take for the Arabs to fight Israel again, war was believed to be all but impossible. It took a full-scale invasion for Israel to reject these faulty beliefs. Following the war, Israel's security and foreign policy shifted dramatically. Prime minister Golda Meir resigned along with much of her cabinet, and both the military chief of staff and the chief of intelligence were dismissed. Not only did Israelis tend to see the war as a disaster (even though they won a military victory on the ground), the Yom Kippur War paved the way to a peace process that Israel would never have considered prior to the war (Johnson and Tierney 2006).

Relying on massive shocks to trigger change in security policy is bad for at least seven reasons. First, it increases the probability of disasters happening in the first place (because the victim fails to act to prevent them). Second, it increases the costs of disasters when they do happen (because the victim is unprepared). Third, it limits future policy options because Congress and/or public opinion disallow similar policies or ventures, even in unrelated contexts (e.g., the "no more Vietnams" rhetoric significantly constrained U.S. military power). Fourth, enemies perceive the victim as vulnerable and ill prepared, encouraging future exploitation or attacks (e.g., 9/11 proved that the U.S. homeland can be struck). Fifth, enemies and allies alike perceive that the victim's deterrence policy has failed, leading them to reconsider their own strategies (e.g., NATO allies were rattled by the Cuban Missile Crisis). Sixth, suffering a disaster compromises a state's credibility, which can demote its effective influence in subsequent international relations (e.g., following the Vietnam war, communists in Southeast Asia could do what they wanted without fear of U.S. intervention, as exemplified by their take over of Cambodia and Laos in 1975). Seventh, the immediate consequences of the disaster give the opponent a first-mover advantage (e.g., the naval losses at Pearl Harbor meant the United States was unable to engage Japanese forces in the Pacific for several months, giving them free reign to conquer the Philippines, Malaya, Hong Kong, Thailand, and numerous Pacific islands, making the Pacific war harder for the United States once it was under way). Any or all of these seven factors can undermine a state's immediate national security, its future influence and power, and the electoral success of its leaders.

### Does It Always Take Disasters to Trigger Change?

Our hypothesis is not that adaptation to novel security threats *only ever* occurs after major disasters, but rather that they often do. But perhaps the United States usually does, in fact, adapt appropriately to new security threats before disaster strikes, and the examples above are merely prominent exceptions to the norm. Further work is needed to provide a comprehensive test of these competing claims. Nevertheless, we offer here a minitest of our hypothesis, as a way of checking how universal the basic problem may be. In order to test the hypothesis that adaptation to novel security threats tends to occur after major disasters, we need an unbiased sample of case studies. For this purpose, we use a list of "watersheds" or turning points in U.S. security policy since World War II, a list that originated in the National Security Department of the U.S. Air War College and has been used in other studies since (True 2002). Table 13.1 lists these cases and, for each, tests the following predictions:

TABLE 13.1. The Seven Post–World War II "Policy Watersheds" in U.S. Security Strategy and Their Conformity to, or Violation of, The Predictions of Our Hypothesis

| Precipitating Event | Predictions | | |
| --- | --- | --- | --- |
| | Disaster? | Unexpected? | Unprepared? |
| Truman Doctrine; Marshall Plan, 1947–1949 [Hogan 1998] | Yes (Iron Curtain falls; spread of communist insurgencies) | **Partially (e.g., Truman doubted implications)** | Yes (massive U.S. policy goals took many years) |
| Rearmament for U.S. containment policy, 1950–1953 [Hastings 1987] | Yes (South Korea invaded) | Yes (as Dean Acheson assured Congress 5 days before the invasion) | Yes (U.S. troops unavailable to assist) |
| Kennedy defense buildup, 1961–1963 [Allison and Zelikow 1999] | Yes (Soviet nuclear missiles in Cuba; West Berlin threatened) | Yes (Kennedy surprised; deterrence strategy failed) | Yes (no plans for a superpower crisis of this type) |

| | | | | |
|---|---|---|---|---|
| Americanization, 1964–1968 [Kaiser 2000] | Vietnam War | Yes (communist expansion in Asia) | **No (but the cost of the war was)** | Yes (badly aligned goals, methods and strategy) |
| Vietnamization, 1969–1973 [Wirtz 1991] | Vietnam War | Yes (Tet offensive in 1968; ultimate defeat) | Yes (Tet was a major intelligence failure; U.S. didn't expect to lose war) | Yes (at Tet troops deployed in wrong places; war strategy misguided) |
| Reagan defense buildup, 1979–1985 [Hayward 2001] | Soviet invasion of Afghanistan | Yes (Soviet expansion) | Yes (full-scale invasion not expected) | Yes (realignment of budget and forces) |
| Reordering of entire U.S. strategic posture, 1990–1991 [Gaddis 1988] | Dissolution of the Soviet Union; Gulf War | **Mixed (collapse of U.S.S.R.; Kuwait invaded)** | Yes (end of Cold War and invasion of Kuwait unexpected) | Yes (U.S. policy changed overnight; Kuwait undefended) |

NOTE: From True 2002. Conformity to predictions is indicated by plain text, and violation of predictions is indicated by boldface text.

1. Disasters tend to precede major changes in security policy.
2. Disasters tend to be unexpected (confirming a failure to foresee it).
3. Disasters tend to be unprepared for (confirming a failure to plan for it).

These predictions are tested against the null hypothesis that the seven policy watersheds resulted from events that were not disasters, and that the United States both expected and was prepared for—in other words, representing a rational, timely adaptation to shifting security threats.

As is clear from Table 13.1, all seven policy watersheds followed dramatic disasters, none of which the United States expected, and for all of which the United States was unprepared. There are just three partial exceptions (boldface text): (1) Postwar Soviet influence in Europe was not entirely unexpected, although the United States and western European allies did not fully recognize Stalin's wider goals until late in World War II. (2) The Vietnam War itself was not unexpected—the United States had already been escalating its commitment under two previous administrations (Eisenhower and Kennedy). Nevertheless, the fighting was far more costly than had been expected. Therefore, the Vietnam War was no less an unexpected disaster than any of the other cases. (3) The collapse of the Soviet Union was a disaster only for the U.S.S.R.; it was the opposite for the United States. However, associated events such as the invasion of Kuwait (along with the spread of civil conflicts in Europe, Asia, and Africa) were very much disasters.

## Why Does It Take Disasters to Trigger Change?

Although states rarely face extinction, their failure to adapt to novel security threats incurs significant costs in blood and treasure. With such a premium on effective adaptation, the pattern of repeated failure in human history begs the question: why does it take disasters to trigger change? It would surely be better to adapt to novel threats incrementally as they arise. Waiting for disasters to happen before adapting begets and worsens those disasters in the first place, and signals weakness to enemies and allies.

Three basic factors impede change. First, change is hard to assess—the consequences are unknown and disasters are rare. Second, change brings uncertainty—if the status quo has worked until now, why risk an uncertain outcome over a familiar one? Third, change entails costs—the reorganization or acquisition of extra resources adds weight to the argument to do nothing.

Beyond these three basic factors, however, a failure to adapt is powerfully exacerbated by converging biological, psychological, organizational, and political phenomena, summarized in Figure 13.1 and explored in detail below.

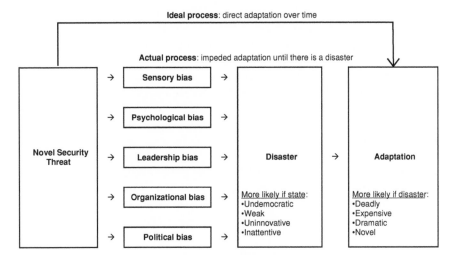

FIGURE 13.1 Scheme of our hypothesis that adaptation to novel security threats tends to occur after major disasters. The causal mechanism is that key biases preserve the status quo and impede adaptation until there is a disaster.

## Sensory Bias

A number of sensory and physiological biases predispose us to maintain the status quo and to avoid expending resources on threats outside our personal realm of experience (see also Blumstein, this volume). Humans have a biological predisposition to react to stimuli that reach our five senses (sight, hearing, taste, smell, and touch), and not to stimuli that remain beyond our personal experience. The machinery of the brain does not fully react to something until we experience it in the flesh. This is unsurprising. Our sensory organs, cognitive architecture, and mental processing evolved in order to respond to real threats and opportunities in our immediate local environment, not to abstract, vague, distant, or hypothetical threats that happen elsewhere, or to others. Of course, our brain does generate vicarious emotional reactions to events that we observe or learn is happening to others, but not as powerfully as if we experience them for ourselves (Simonsohn et al. 2006). Such effects are evident in international relations as well. Decisions about military intervention were found to be influenced more by a state's own experience than merely observing the experience of others (Levite et al. 1992). The United States, for example, was unperturbed about the French experience of war in Vietnam—Kennedy reminded a reporter, "That was the *French*. They were fighting for a colony, for an ignoble cause. We're fighting for freedom" (Tuchman 1984, 287). Later on, the Joint Chiefs of Staff, citing French errors and indecision in the conflict, noted wryly, "The French also tried to build the Panama Canal" (U.S. Department of Defense 1971, Vol. 3, 625).

Also very important is the general principle, across a wide range of psychological phenomena, that negative events and information are processed more thoroughly and have greater impact than positive events, and negative impressions and stereotypes are quicker to form and more resistant to disconfirmation than positive ones (Baumeister et al. 2001). In terms of the effects of experience on human psychology, "bad is stronger than good." In international politics as well, failure, as opposed to success, appears to have an intrinsic leverage: "People learn more from failure than from success . . . past success contributes to policy continuity whereas failure leads to policy change" (Levy 1994, 304). This appears to result from an interaction with expectations. "Outcomes that are consistent with expectations and achieve one's goals generate few incentives for a change in beliefs, whereas unexpected results and those that fall short of one's goals are more likely to trigger a change in beliefs and policy. Thus the most likely outcomes to trigger learning are failures that were either unexpected at the time or unpredictable in retrospect" (Levy 1994, 305). A classic study by Dan Reiter found that alliance behavior was most influenced by a state's experience of success or failure in previous wars, and ignored actual current threats (Reiter 1996). States switched their policy only if it was deemed a failure in the past.

In summary, we are most likely to react to a threat (1) if it reaches us through first-person experience (rather than via newspapers, radio, the Internet, or television), and (2) if it is a negative event (such as a disaster) rather than a positive one. It may therefore take a Pearl Harbor of 1941, a threat of nuclear holocaust as in 1962, defeat in war, or a 9/11 to surmount our sensory barriers, acknowledge major new threats, and goad us into action.

## Psychological Bias

A number of psychological biases also predispose us to maintain the status quo and to avoid expending resources on threats outside our normal realm of perception. Perhaps most important is cognitive dissonance. Conflicting information must be resolved in order to generate a coherent interpretation, and cognitive dissonance tends to select, organize, or distort incoming information so that it matches our preferred or preexisting beliefs (Vertzberger 1990; Tetlock 1998; Sears et al. 2003; McDermott 2004). Even experts often discount potential problems due to the cognitive demands of complex events (Dorner 1996). For example, Irmtraud Gallhofer and Willem Saris found that despite at least seven distinct strategies being floated during the Cuban Missile Crisis executive committee meetings, decision makers tended to consider only two at a time (Gallhofer and Saris 1996).

Experimental research in cognitive and motivational psychology reveals a vast array of biases that tend to preserve the status quo: *deformation professionelle* (a tendency to see things from the perspective of the conventions of one's profession); the mere exposure effect (a preference for things that are more familiar); the availability heuristic (a tendency to make predictions that are based on perceived rather than actual salience); projection bias (a tendency to assume that others share similar beliefs to oneself); the bandwagon effect (a tendency to do or believe the same as others); false consensus effect (a tendency to expect others to agree with oneself); discounting (to prefer immediate over long-term payoffs); and, finally, the well-documented and pervasive effects of in-group favoritism and out-group derogation, groupthink, and overconfidence (Janis 1972; Jervis 1976; Kahneman et al. 1982; Vertzberger 1990; Tetlock 1998; Johnson 2004).

Overconfidence appears to have a particular importance. We tend to hold positive illusions of our abilities, our control over events, and of the future, all of which lead to overconfidence about our vulnerability to risk, and therefore to discount the need for change (Johnson 2004). Positive illusions in U.S. decision making may account for the failure to deter Japan in 1941, the Soviet Union in 1962, and Saddam Hussein in 2003, among other cases. However, harking back to the importance of sensory biases, once personally involved in a disaster, optimistic illusions disappear. Psychologists found that Californians were overly optimistic about the risk of earthquakes until they lived through one (Burger and Palmer 1992). Yechiel Klar's study of Israelis living with the threat of terrorist attacks found that people maintain positive illusions as long as threats are "hypothetical" and "psychologically unreal." But, "when the group to which people belong is the target of some significant ongoing calamity, even when the participants themselves are currently not the direct victims, the unreality of the event dissolves and optimism (both absolute and comparative) decreases or vanishes altogether" (Klar et al. 2002, 216). Disasters serve to wake us up to reality. They are very effective at doing so, but, by definition, the wake up call comes too late.

## Leadership Bias

Particular leaders and their ideas often compel us to maintain the status quo and to avoid expending resources on threats outside the accepted realm of attention. These individuals' preferences can also become institutionalized such that they persist beyond their worth until, or sometimes even after, the original proponent falls from power, resigns, or dies. It has been a recurrent historical theme for leaders to derail their own intelligence services by favoring positive reports, punishing the bearers of bad news, setting different agencies in competition with each other, and interfering with the methods and targets of information gathering (Handel 1989; Van Evera 2003). A recent example is the Bush administration's use of intelligence on weapons of mass destruction and the postwar challenges of Iraq in order to support their favored policy (Clark 2003; Jervis 2003; Fallows 2004; Woodward 2005). With such strong incentives to control information and policy, and to protect their political reputation, leaders can exert enormous impediments to effective adaptation.

## Organizational Bias

Numerous organizational biases also predispose us to maintain the status quo and to avoid expending resources on threats outside the realm of standard operating procedures. Bureaucratic procedures, vested interests, competition for promotions, sunk costs, access to the elite, and turf wars over budgets and responsibilities favor a rigid focus on past events and successes, and a rigid avoidance of rocking the boat to advocate some new and unproven revision of strategy (Kovacs 1997; Allison and Zelikow 1999; Van Evera 2003). An entire literature has built up around this principle (organizational learning) and forms the classic "bureaucratic politics model" of decision making in political science—a default explanation for bizarre or failed policies (Allison and Zelikow 1999). Although organizational biases may create problems, these very characteristics are to some extent intentional: "Indeed, the value of institutions typically lies in their persistence or 'stickiness,' which allows actors to make plans, invest and organize their affairs around institutions and, in general, lends certainty and predictability to their interactions" (Viola and Snidal 2006, 5).

At times, however, the costs will outweigh the benefits. Prior to 9/11, the machinery, professionals, and mindsets of the Cold War era still exerted a significant legacy. There was a "failure of imagination"—a dearth of lateral thinking or fresh ideas—in the intelligence community even though the threats of transnational terrorism were evident (Simon and Benjamin 2000; Rosenau 2007). In addition to the failures to actually plan for novel threats, Stephen Van Evera has laid out reasons why institutions have little incentive to self-criticize or evaluate their own performance at all (Van Evera 2003). The entire institutional environment is hostile to adaptation: "Myths, false propaganda, and anachronistic beliefs persist in the absence of strong evaluative institutions to test ideas against logic and evidence, weeding out those that fail" (Van Evera 2003, 163). The maintenance of bureaucracy itself can sometimes become an all engrossing task. When Donald Rumsfeld took over at the Pentagon, for example, he began issuing numerous white memos ("snowflakes") around the Pentagon demanding information on who actually does what and how they do it, until some people were spending more time answering snowflakes than doing normal work (Woodward 2005).

A further problem with organizations is that the "sensors"—the people with their ears to the ground—are disjointed from the decision-making structure (in an interesting corollary to the sensory failures noted above). Leaders are sometimes the last to know about impending (or even actual) disasters. The middle managers or those below them are the ones who deal on an everyday basis with the outside world and are therefore more likely to detect novel threats, or to recognize that old methods are no longer appropriate. For example, when a minor flaw was found in the Pentium processor in 1994, Intel suffered half a billion dollars of damage in under six weeks. The fault caused a rounding error in division just once every nine billion times, however, this tiny flaw quickly became significant—the news spread rapidly on the Internet and was amplified by Intel's new global prominence and identity. According to Intel CEO Andrew Grove, "I was one of the last to understand the implications of the Pentium crisis. It took a barrage of relentless criticism to make me realize that something had changed—and that we needed to adapt to the new environment" (Grove 1999, 22).

This echoes the intelligence situation before 9/11 and the reaction of administration officials. CIA director George Tenet and terrorism expert Cofer Black say they could not have laid out the serious possibility of a major attack on U.S. soil any clearer to Condi Rice in a meeting in July 2001. "The only thing we didn't do," according to Black, "was pull the trigger to the gun we were holding to her head" (Woodward 2005, 79). If the National Security Adviser is unreceptive to such an issue, then it is unlikely to win the President's attention. The administration as a whole was simply not geared to respond to the growing threat of al-Qaeda (Gellman 2002; Clarke 2004). Even if they had been receptive, as one insider noted, "The U.S. government can only manage at the highest level a certain number of issues at one time—two or three. You can't get to the principals on any other issue" (Gellman 2002).

Organizational and bureaucratic impediments to security appear to be severe. Eventually, budgets or political obstacles get in the way. Richard Betts's analysis of surprise attacks in international relations found that "most of the options available to the West for reducing vulnerability to surprise are limited by political or financial constraints" (Betts 1983, 311). Effective readiness against major threats was simply too expensive or complicated to maintain on a regular basis.

## Political Bias

Electoral politics also predispose us to maintain the status quo and to discount genuinely important threats in favor of politically salient ones. There is no reason to expect efficient adaptation (or sometimes any adaptation at all) to address the most important national security threats. What is threatening in secret intelligence reports is irrelevant to an oblivious public—or rather, an oblivious electorate. Politics provides numerous alternative motivations for individual leaders, political parties, lobby groups, and the public to steer policy and incentives in their own preferred direction, often to the detriment of adaptation to national security threats. The reality of politics means that radical shifts in policy, especially toward a novel hypothetical threat (about which the key intelligence information may be known only to elites) are often indefensible in Congress, hard to obtain the necessary budget to initiate or complete, and politically suicidal. There are few points to be scored (or as many to lose) in pushing for rapid or comprehensive change, for admitting mistakes, or for adapting. As long as the threat is at least four years away, or can be blamed on extraneous causes or opposing political parties, other concerns are likely to take precedence.

Incumbency is an important component of this problem. A high turnover of civil servants or politicians allows for continual and gradual adaptation to changing circumstances over time. By contrast, a low turnover reduces the ability and inclination to adapt, gradually bottling up problems until the whole system collapses under the pressure of a major disaster. In the U.S. government, a number of factors operate to empower incumbents and entrench particular elites and procedures. Disasters may be particularly effective at bringing down an incumbent regime, whose failings—real or perceived—often become a central motivation and electoral strategy for opposition parties or congressional inquisitions.

To summarize this section, numerous features of human nature and the nature of institutions that humans create limit our ability to detect and react appropriately to novel security threats. Because these features stem from independent sources at different levels of analysis (e.g., individual behavior, organizational behavior, elite decision making, etc.), they are likely to generate a status quo bias across a wide range of circumstances. For example, even a forward-looking bureaucracy may have to work against a short-sighted leadership, or vice versa. To put it bluntly, society seems predisposed to preserve the status quo until something goes wrong. As Henry Petroski noted in his book *Success through Failure*, "Good design always takes failure into account and strives to minimize it. But designers are human beings first and as such are individually and collectively subject to all the failings of the species, including complacency, overconfidence, and unwarranted optimism" (Petroski 2006, 193–194).

## When Are Disasters More or Less Likely to Trigger Change?

*When* are states more likely to suffer disasters? And what *types* of disasters are more or less likely to generate appropriate and lasting change? In other words, what are broad-brush circumstances (or "independent variables") that work to exacerbate or suppress the biases we have noted above? Such sources of variation are crucial to future tests of our hypothesis. But they also have practical significance: if we can get a handle on the basic conditions that make disasters more or less likely, we can attempt to steer our behavior and institutions toward those conditions that reduce the probability of being the victim of disaster. Below we consider characteristics (of both states and disasters) that are most and least likely to cause adaptation to novel security threats.

## Characteristics of States That Promote Adaptation

*Democracy.*  Democracies promote journalistic inquiry, congressional review, opposition criticism, and the regular turnover of political representatives. By contrast, authoritarian regimes deter or silence messengers of bad news and favor long-term incumbents. Hitler's generals, for example, rarely told him the truth about the impending disasters such as at Stalingrad (Handel 1989; Beevor 1998). The influence of democracy is a matter of degree, however, rather than just a binary distinction between democracy and tyranny. For example, the Bush administration's handling of intelligence prior to the Iraq War served to undermine the Washington system of checks and balances, handing the authority to wage war to the President— exactly what the founding fathers designed the U.S. government to avoid (Fisher 2003).

*Power.*  Powerful states can create expensive and extensive intelligence agencies, equipment, and personnel. By contrast, weak states are more likely to be constrained by the resources they have to detect, prepare for, and react to disaster. Of course, 9/11 and many other examples given above demonstrate that even massive amounts of resources available to powerful states such as the United States do not solve the problem. Power must be applied effectively. Nevertheless, on average, more powerful states should be more likely to achieve effective adaptation.

*Innovation.*  Security strategies adapt more effectively in a more innovative culture. For example, Germany and Britain were far more innovative in developing their military strategies and tactics than the French in the interwar period, and this directly led to differential combat outcomes in World War II (Posen 1984).

*Attention.*  When a state is focused and committed to dealing with a potential threat, it has a much higher chance of adapting to meet it. By contrast, when a state is mired in serious domestic or international crises, it is far less likely to detect or respond appropriately to novel threats.

## Characteristics of Disasters That Promote Adaptation

*Deadly.* High numbers of casualties, especially civilians.
*Expensive.* High levels of damage (or lost opportunity).
*Dramatic.* Symbolic or salient targets.
*Novel.* Not just a big version of an existing threat.

## Can Change Occur without Disaster?

Our hypothesis is that adaptation to novel security threats tends to occur after major disasters. Like any hypothesis, this does not mean that policy changes *only ever* occur following major disasters. We only suggest that it may be more commonly the case than the other way around. But there are likely to be exceptions. Indeed, our discussion above sets out explicit conditions under which we may expect major policy change to occur without disaster—powerful, democratic states that are innovative and attentive, especially ones with minimal biases in their psychology, organizational behavior, leadership, and politics. Therefore, not only do we expect there will be exceptions, but we propose specific variables that will characterize these exceptions. Future studies could test this hypothesis with, for example, matched pairs of otherwise similar cases: one that adapted successfully, and one that did not.

One can think of a number of potential counterexamples in which security strategy changed significantly without any disaster. For example, people often cite the remarkable lack of a disaster following the break up of the Soviet Union and the democratization of Eastern Europe after the fall of the Berlin Wall. However, it is important to realize that this is a distinctly western perspective. From the perspective of the U.S.S.R., it was the biggest disaster of its history—indeed, it signaled its own extinction. Even from a western perspective, though, the lack of disaster may be misleading: the new security environment represented the collapse of a formidable enemy, followed by a power vacuum, not the emergence of a new threat per se. There was no agent to bring disaster (barring a renegade general with Soviet nuclear missiles, or something along those lines). Similar problems arise with many other prominent examples of major security changes that appeared to escape disaster: the fall of the British Empire, the reunification of Germany, the nuclear armament of India and Pakistan. Such cases deserve further scrutiny in the context of our hypothesis.

It is easier to think of counterexamples that do not reside in the realm of security. For example, numerous social, economic, and technological transformations occur without disaster, ranging from awarding women the right to vote, to the introduction of the Euro, to the space race. One could also argue that major environmental efforts are underway to avert the looming disaster of global climate change (as evidenced by such forward thinking as the Stern Review). However, it is not at all clear that anywhere near enough is actually being done, whatever the proposals and plans.

Perhaps, then, there is something special about the domain of security that links change and disaster. After all, lapses in security are almost by definition associated with violence, death, and destruction, so security disasters may be more dramatic, more visible, and more likely to compel policy makers and organizations to change.

Even if adaptation is rare in the domain of security, future studies can identify cases where adaptation was more successful than others. One example might be the Malaya Emergency of 1948–1960 (see Johnson and J. Madin, this volume). In that conflict, the British and Malayan counterinsurgency forces revealed themselves as organizations able to learn and adapt—though tellingly this occurred *through* a series of disasters. According to a recent survey, "one of the things that allowed the British army to innovate and adapt during its counterinsurgency operations in Malaya in the 1950s (and thus attain success) was its willingness at all levels to admit failure" (Metz and Millen 2004, 26; Nagl 2002). The key comparison may therefore be between states that do learn from disasters, and states that do not learn even from disasters. On that note, we now turn to possible solutions to minimize the likelihood of disasters.

## Solutions

The bright side of this story is that the reasons for our failure to adapt are systematic, not random. Empirical evidence from cognitive and social psychology offers a taxonomy of causes and consequences of key biases (reviewed in Jervis 1976; Tetlock 1998; Sears et al. 2003; Van Evera 2003). We therefore have scientific tools to identify when, where, and why we fail to adapt to new threats, who is most susceptible, and how to make corrections to compensate (or even overcorrections as insurance policies against these biases). It is likely, however, to be a difficult task—we are often blissfully unaware of these biases in the first place, and that is precisely why we fall prey to their influence. Moreover, solutions require us to look beyond our typical experience and plan for things that seem unlikely and farfetched—hardly things that motivate urgent action. Nevertheless, a careful study of causes and consequences can, in principle, help to design institutions and decision-making procedures that will improve adaptation to novel security threats.

Some of the problems outlined above are already well recognized by the policy community. Indeed, many of the key problems are already being addressed by the post-9/11 reorganization of the intelligence services. For example, the U.S. National Intelligence Council was set up to look ahead at emerging threats that remain "over the horizon." Other changes include more scenario-based planning exercises, more "red teaming" (role-playing the enemy), and recruiting lateral and imaginative thinkers such as the novelist Tom Clancy to think through possible future threats. However, reforms may be more successful if they exploit the scientific insights emerging from biology and psychology. If history is any guide, the same mistakes will be repeated unless we try something new. We need innovative solutions if we are to escape the recurrent failures of imagination that litter the past so liberally. We discuss a number of potential solutions below.

## Lessons from Evolutionary Biology

The best model of successful adaptation to changing security threats may lie in evolutionary biology, where adaptation is the core process underlying billions of years and millions of examples of survivors. Adaptive processes in nature are magnificently diverse, fine tuned over countless generations of trial and error, and well documented. In his analysis of security insights from biological evolution, Vermeij (this volume) notes seven key strategies that can be employed in the face of novel security threats: tolerance, active engagement, increase in power or lifespan, unpredictable behavior, quarantine and starvation of the threatening agent, redundancy, and adaptability. He finds that "the most successful attributes of life's organization—redundancy, flexibility, and diffuse control—are also the characteristics of human social, economic, and political structure that are best suited to cope with unpredictable challenges." We list Vermeij's conclusions in Table 13.2, along with some suggested applications to security.

Vermeij's key insight is that adaptations to everyday threats often also turn out to be effective adaptations to unpredictable threats. The owners of these serendipitous adaptations will be more likely to avoid or withstand rare and unpredictable disasters. As Vermeij puts it, a trait that enabled an organism to "endure the extraordinary conditions prevalent during times of mass extinction cannot be considered an adaptation to those circumstances but is instead an accidental if welcome consequence of adaptation to more commonplace phenomena" (Vermeij, this volume). There are numerous examples in human history in which commonplace adaptations were used to deal with novel threats. For example, when Soviet tanks escorting convoys in Afghanistan discovered they could not elevate their guns high enough to engage hostile forces high on the mountainsides, the Soviet Army resorted to using self-propelled anti-aircraft artillery instead (Beckett 2001). States that accumulate diverse and flexible technologies, practices, or institutions over time are more likely to be able to fall back on a broader range of alternatives in unusual circumstances.

TABLE 13.2. Vermeij's Lessons from the History of How Biological Organisms Evolved to Deal with Unpredictable Threats in Nature (Vermeij, this Volume), and Some Possible Applications to Security Policy Derived from Our Study

| Lessons | Application |
|---|---|
| There will always be unpredictable threats, and no adaptations to them can ever be perfect. | We should expect ongoing arms races rather than perfect solutions. Even imperfect adaptations lead to improved strategies. |
| Adaptation to threats comes with costs and constraints. | Adaptation may be costly, but stasis may be worse. Adaptation must be allocated rolling budgets (not one-off lump sums). |
| Passive resistance, though highly effective, is inconsistent with activity and the exercise of power and by itself is not an acceptable option for most human societies. | Proactive strategies are essential if a state wants to play other international roles. U.S. isolationism is inconsistent with its defense. |
| Exclusively active engagement exposes entities to ecological collapse engendered by interruptions in resource supplies and, therefore, by itself is an unreliable long-term strategy. | Unlimited commitment to active engagement is risky and may be counterproductive. |
| Redundancy and a modular structure of semiautonomous parts under weak central control provide the most flexible, adaptable, and reliable means of making unpredictable challenges predictable. | Policymaking, military, and intelligence resources should be decentralized, granted independence, and have back-up systems. |
| The history of life in general, and of extinction in particular, shows that adaptation to everyday as well as unpredictable circumstances has improved over the course of time. | Adaptations that can be co-opted to alternative uses offer dual protection against commonplace and unpredictable threats. |

## Lessons from the Immune System

Immune systems also offer intriguing models for human security. They are especially interesting because of their efficiency, lying low in normal times but wielding an extraordinary capacity for an enormous surge in response to a threat. As Villarreal (this volume) writes, "Biological systems are inherently local, rapid, robust and adaptable systems. They are able to rapidly marshal all the needed diverse and central resources, but inherently reduce resource consumption when no longer needed. They are capable of searching for, finding, destroying, and sterilizing threats, both hidden and apparent. They are even able to respond to threats never before seen." The prominence of immune responses in nature attests to the advantages of flexibility and adaptability in the face of novel threats. However, the immunity model presents an additional point: locally focused responses may be far more adept at contending with new threats than those requiring central control or approval. This has potential implications for security strategy in human systems; central command and control structures are often less able to detect, understand, and respond adequately to new threats than local organizations in direct and immediate contact with the threat. Interestingly, there are cases in which the immune system can overreact, drain significant resources, and become dangerous to the organism itself. This also has parallels in human security, in which perceived threats can initiate overblown and costly responses (Mueller 2005; Blumstein, this volume).

## Lessons from Institutional Design

Institutions and organizations could be redesigned to hard wire mechanisms for effective adaptation, just as DNA and the process of natural selection assure adaptation in biology. A recent study by Viola and Snidal (2006) argues that evolutionary mechanisms offer a "potentially powerful way to account for the persistence, adaptation, and abandonment of international institutions" (Viola and Snidal 2006, 3). Although current international institutions exhibit many features that arose by design, they also exhibit many other features that arose from a process of "decentralized emergence" over time, without conscious planning. For example, norms of sovereignty, diplomacy, and customary international law arose largely "from the on-going practices of states" (Viola and Snidal 2006, 1). Ideally, institutions would include such organic attributes in order to "adapt and respond to unanticipated elements in their environment" (Viola and Snidal 2006, 4). Designers could identify how these processes of adaptation occur, the conditions under which they are successful, and ways to exploit them. Of course, adaptability often already exists: some degree of flexibility is granted in most organizations, decisions may be delegated to lower level units, and mechanisms are often in place to seek and respond to feedback. Nevertheless, an evolutionary approach may help to identify successful adaptive processes, their likely causes, and their likely consequences.

In addition, the methods and quantitative tools developed in biology to study adaptation may prove useful in understanding the adaptation of human institutions as well. Viola and Snidal note that "it is unlikely that institutions would develop without growing in some way out of the previous institutions," and "in a given issue area it is common to see institutions with family resemblances" (Viola and Snidal 2006, 8). These echo the notions of common ancestry and evolutionary legacy central to evolutionary biology, for which there are well-developed statistical methods to test for evidence of adaptation, correlations with associated traits, and points of divergence, all while controlling for characteristics shared by common ancestry.

In a fast-moving world of rapid communication, some have argued that even forward planning is no longer the best strategy to prepare for the future. Instead, organizations can be structured to be automatically adaptable and flexible by nature, so that the system is self-geared to adapt and exploit change as it happens (Brown and Eisenhardt 1998). This violates many traditional views of organizational design, but at least one prominent firm is based on this kind of unstructured system: Google (Lashinsky 2006). Google actively promotes innovation and experimentation through the independence of its subunits and workers. One strategy is encouraging its engineers to spend one day a week working on pet projects and submitting new ideas for product development to the Google "ideas list" (Elgin 2005). This list is monitored by the upper levels of management (as opposed to first passing through multiple middle levels) and screened for highly innovative and potentially investment-worthy ideas.

Finally, even if an organization itself cannot easily be restructured, incentive structures can be created within it (via financial, budgetary, or professional rewards) to encourage flexibility, adaptation, and review instead of rigidity, policy stasis, and nonevaluation.

## Lessons from the Insurance Industry

In order to remain financially viable, insurance companies must be able to either predict or build in buffers against novel catastrophes. The insurance industry thus provides another interesting model for contending with future threats—both known and unknown. In a sense, these companies provide a form of "preadaptation" to novel threats—a guarantee of being able to rebuild following damage. Although it is inherently costly, adopting insurance strategies can provide the necessary buffers against occasional disasters. Although the disasters themselves may not be possible to avoid, their negative consequences can be mitigated. However, it is important to recognize that insurance companies have the luxury of passing on these costs to their clients; government agencies do not.

## Lessons from Futures Markets

If humans are bad at detecting novel threats, an alternative is to maximize the number of individuals contributing to assessment. It is a well-recognized phenomenon that the average of a large number of estimates can be extremely accurate—the so called "wisdom of crowds" (Surowiecki

2004). As long as the group is diverse, independent, and decentralized, then individual biases will cancel each other out, leaving available information from a wide range of sources to converge on the correct assessment. For exactly this reason, even expert analyses by intelligence agencies, such as the CIA, may be expected to be inaccurate, because they are not diverse (analysts are all Americans), not independent (analysts share methods, sources, and information), and not decentralized (they work for the same organization). By contrast, an ideal assessment would include opinions from across the globe, including the full spectrum of ideological, cultural, and political differences, and exploiting multiple sources of local information. This has direct practical applications. Harnessing this phenomenon and using it for predictions can be achieved by the use of "futures markets," in which one buys a contract that will pay, say, $10 if a given event occurs by a certain date. The market price of these futures contracts then reveals a probability that the event will happen (Leigh et al. 2003). For example, on February 14, 2003, the price of $10 futures on Saddam Hussein no longer being president of Iraq on June 30 were trading at $7.50 on tradesports.com, suggesting the probability of war was 0.75.

The Pentagon proposed a "Policy Analysis Market" to exploit the opportunities of futures prediction in 2003. The idea was to use futures markets to evaluate growth, political stability, and military activity in eight nations, four times a year. The project swiftly attracted the misnomer of "terrorism futures" and was scrapped by nervous politicians—yet another example of institutional bias working against innovation. Nevertheless, a number of political futures markets do exist on commercial web sites. One can bet, for example, on the likelihood of U.S. military action against North Korea, air strikes against Iran, or the capture of Osama bin Laden. If these futures markets can be expanded, they may well outperform expert assessments of the likelihood of important events in national security, bypassing the impediments and biases to adaptation outlined above.

## Conclusions

If humans, institutions, and states were rational, security policy would change in step with the shifting threats of the day. Our examples of Pearl Harbor, Cuba, Vietnam, and 9/11 indicate that this logic is often violated, and the United States failed to adapt to novel security threats until they caused a major disaster. Our mini case study suggests that these examples are not unusual (Table 13.1). On the contrary, all seven U.S. security policy "watersheds" since World War II were initiated by major disasters, which the United States neither expected nor prepared for. This is further supported by the fact that the U.S. defense budget has not changed in line with shifting threats but rather as significant step-changes after major international events (True 2002). Adaptation to novel security threats is most likely to occur when a state suffers a major disaster—especially among states that are democratic, powerful, innovative, and attentive, and especially if the disaster is deadly, expensive, dramatic, and novel. In other, "normal" times, adaptation to novel security threats is severely impeded because (1) dangers that remain hypothetical fail to trigger appropriate sensory responses, (2) psychological biases serve to maintain the status quo, (3) dominant leaders entrench their own idiosyncratic policy preferences, (4) organizational behavior and bureaucratic processes resist change, and (5) electoral politics offers little incentive for expensive and disruptive preparation for unlikely and often invisible threats. The sudden disasters that break intervening periods of stasis are analogous to the paradigm shifts that Thomas Kuhn (1970) noted in the progress of science, and the punctuated equilibrium theory that Frank Baumgartner and Bryan Jones (1993, 2002) proposed to explain the dynamics of U.S. policy making.

Even when adaptations do follow disasters, they often turn out to be short-lived. Soon enough the powerful impediments to change, whether psychological, organizational, or political, come to the fore. The human brain tends to cast our perception of past events in an overly positive light (Greenwald 1980; Schacter 1995). Even after the unprecedented carnage of World War I, for example, John Stoessinger noted that the "old people to whom I spoke about the war remembered its outbreak as a time of glory and rejoicing. Distance had romanticized their memories, muted the anguish, and subdued the horror" (Stoessinger 1998, xii). Organizations and societies also work to downplay failure and construct myths that deflect blame and reinterpret history (Van Evera 1998; Schivelbusch 2004; Johnson and Tierney 2006). For example, German society embraced the myth after World War I that the army was undefeated on the battlefield and had been stabbed in the back by politicians. Meanwhile, political elites go through the motions, creating the image of change without any intention of bearing its real costs, or doing just enough to tick the boxes in the eyes of Congress or the public. Even with 9/11, for example, the disaster appears to have paled enough into the past that essential reforms have fallen far below the recommendations of the 9/11 commission (9/11 Public Discourse Project 2005). A similar process occurred after the bombing of the London Underground: "The atrocities of July 7th 2005 turn out to have been the kind of alarm call that is followed by intemperate grunts and a collective reaching for the snooze button" (Economist 2006). It is noticeable that political, media, and public attention has already strayed from terrorism and the war on terror in favor of a new sensory-rich disaster on which everyone is focused: Iraq.

Fortunately, cumulative major disasters such as 9/11 usually generate a kind of ratchet effect, such that even after the initial impact wears off, we

TABLE 13.3.  Policy Prescriptions to Maximize Effective Adaptation
in Each of the Key Problem Areas

| Bias | Policy Prescriptions |
|---|---|
| Sensory bias | Ensure decision makers see frontline personnel and victims |
| | Ensure decision makers hear opposing viewpoints |
| | Ensure decision makers travel to places at issue |
| Psychological bias | Increase diversity and sources of information |
| | Increase turnover in appointees and decision-making groups |
| | Install high-level devil's advocates in policy discussions |
| Leadership bias | Limit power |
| | Limit terms of office |
| | Insist on periodic reevaluation of existing policies |
| Organizational bias | Encourage "bottom-up" development and communication of ideas (Google model) |
| | Solicit recurring internal and external review |
| | Create incentives for continual change |
| Political bias | Increase public information (so that electorate and government see the same threats) |
| | Increase congressional oversight of security policy |
| | Reduce campaign financing and duration |

are still left with some, perhaps imperfect, novel adaptations (e.g., improved airport security, or the U.K. Civil Contingencies Secretariat to "prepare for, respond to and recover from emergencies," see www.ukresilience.info). It is often noted that in Chinese, the word for "crisis" includes the notion of opportunity as well as danger. If humans are not good at avoiding disasters, we should at least learn to react to them in ways that best utilize the opportunity for change. Cumulative change can be maximized even if it is frustratingly imperfect.

Democratic, powerful, innovative, and attentive states may have the best chance of avoiding security disasters. But whether a state meets these conditions or not, there are a number of policy prescriptions that could improve effective adaptation to novel security threats (Table 13.3). Future studies will be able to improve, expand, and test these ideas, and there is clearly a wealth of models from which to derive effective tricks of adaptation, including evolutionary biology, the immune system, institutional design, futures markets, and insurance.

Although there is room for improvement, history suggests that humans need disasters to occur before waking up to novel security threats, whether they are disasters of national security, disease, starvation, poverty, or

---

## Hypothesis and Predictions

### HYPOTHESIS
Adaptation to novel security threats tends to occur after major disasters.

### PREDICTIONS
Disasters tend to precede major changes in security policy.
Disasters tend to be unexpected (confirming a failure to foresee it).
Disasters tend to be unprepared for (confirming a failure to plan for it).

---

environmental change. This does not bode well for the future. Even when a threat poses a clear and present danger, such as global climate change, political actors do almost nothing to adapt to the threat until it is too late. As a recent *New Scientist* editorial recognized (New Scientist 2006): "The world will one day act with urgency to curb greenhouse gases: the likely violence of the atmosphere's reaction to our emissions makes that inevitable. Climate change awaits its 9/11."

### ACKNOWLEDGMENTS
We thank Rafe Sagarin and Terence Taylor for their ideas, advice, criticism, and invitation to join the Working Group on Ecological and Evolutionary Models for Homeland Security Strategy; and the National Center for Ecological Analysis and Synthesis for hosting us. Dominic Johnson is indebted to the Branco Weiss Society in Science Fellowship, the International Institute at UCLA, and the Society of Fellows and the Woodrow Wilson School of Public and International Affairs at Princeton University. Elizabeth Madin would like to thank the U.S. Department of Homeland Security, the U.S. National Science Foundation, and Steve Gaines. We also owe our thanks to Richard Cowen, Robert Mandel, Dominic Tierney, and all the members of the working group for excellent comments and criticisms on the manuscript.

### REFERENCES
9/11 Commission. 2004. *The 9/11 Commission report: Final report of the National Commission on Terrorist Attacks upon the United States.* New York: W. W. Norton.

9/11 Public Discourse Project. *2005 final report on 9/11 Commission recommendation.* December 5. Available at www.9-11pdp.org/.

Allison, G., and P. Zelikow. 1999. *Essence of decision: Explaining the Cuban missile crisis.* New York: Longman.

Baumeister, R. F., E. Bratslavsky, C. Finkenauer, and K. D. Vohs. 2001. Bad is stronger than good. *Review of General Psychology* 5: 323–370.

Baumgartner, F. R., and B. D. Jones. 1993. *Agendas and instability in American politics*. Chicago: University of Chicago Press.

Baumgartner, F. R., and B. D. Jones. 2002. *Policy dynamics*. Chicago: University of Chicago Press.

Beckett, I. F. W. 2001. *Modern insurgencies and counter-insurgencies: Guerrillas and their opponents since 1750*. New York: Routledge.

Beevor, A. 1998. *Stalingrad*. London: Penguin.

Betts, R. 1983. *Surprise attack: Lessons for defense planning*. Washington, DC: Brookings Institution Press.

Blum, H. 2003. *The eve of destruction: The untold story of the Yom Kippur War*. New York: HarperCollins.

Brown, S. L., and K. M. Eisenhardt. 1998. *Competing on the edge: Strategy as structured chaos*. Cambridge, MA: Harvard Business School Press.

Burger, J., and M. Palmer. 1992. Changes in and generalization of unrealistic optimism following experiences with stressful events: Reactions to the 1989 California earthquake. *Personality and Social Psychology Bulletin* 18: 39–43.

Busenberg, G. J. 2003. Agenda setting and policy evolution: Theories and applications. Paper presented at The Midwest Political Science Association 2003 Conference. Chicago, IL, April 3–6.

Clark, W. K. 2003. *Winning modern wars: Iraq, terrorism, and the American empire*. New York: PublicAffairs.

Clarke, R. A. 2004. *Against all enemies: Inside America's war on terror*. New York: Free Press.

Cohen, E. A., and J. Gooch. 1991. *Military misfortunes: The anatomy of failure in war*. New York: Vintage.

David, S. 1997. *Military blunders: The how and why of military failure*. London: Robinson.

Dixon, N. 1976. *On the psychology of military incompetence*. London: Jonathan Cape.

Dorner, D. 1996. *The logic of failure: Recognizing and avoiding error in complex situations*. Cambridge, MA: Perseus.

Economist. 2006. One year on: The wake-up call that wasn't. *Economist*, July 8, 29–30.

Elgin, B. 2005. Managing Google's idea factory. *Business Week Online*, October 3. http://www.businessweek.com/magazine/content/05_40/b3953093.

Fallows, J. 2004. Blind into Baghdad. *Atlantic Monthly*, January/February, 53–74.

Fisher, L. 2003. Deciding on war against Iraq: Institutional failures. *Political Science Quarterly* 118: 389–410.

Foucault, M. 1970. *The order of things: An archaeology of the human sciences*. New York: Random House.

Foucault, M. 1977. Nietzsche, geneology, history. In *Language, counter-memory, practice: selected essays and interviews*, ed. D. F. Bouchard, 139–164. Ithaca, NY: Cornell University Press.

Fursenko, A., and T. Naftali. 1997. *One Hell of a gamble: Khrushchev, Castro, and Kennedy, 1958–1964*. New York: W. W. Norton.

Gabriel, R. 1986. *Military incompetence: Why the American military doesn't win*. New York: Noonday Press.

Gaddis, J. L. 1988. *We now know: Rethinking cold war history*. New York: Oxford University Press.

Gallhofer, I. N., and W. E. Saris. 1996. *Foreign policy decision-making: A qualitative and quantitative analysis of political argumentation*. Westport, CT: Praeger.

Gartner, S. S. 1997. *Strategic assessment in war.* New Haven, CT: Yale University Press.

Gellman, B. 2002. A strategy's cautious evolution. *Washington Post,* January 20.

Gilbert, M. J. 2002. *Why the North won the Vietnam War.* New York: Palgrave.

Greenwald, A. G. 1980. The totalitarian ego: Fabrication and revision of personal history. *American Psychologist* 35: 603–618.

Grove, A. S. 1999. *Only the paranoid survive: How to exploit the crisis points that challenge every company.* New York: Currency.

Handel, M. I., ed. 1989. *Leaders and intelligence.* London: Frank Cass and Co.

Hastings, M. 1987. *The Korean War.* London: M. Joseph.

Hayward, S. F. 2001, *The age of Reagan: The fall of the old liberal order.* Roseville, CA: Prima.

Hogan, M. J. 1998. *A cross of iron: Harry S. Truman and the origins of the national security state, 1945–1954.* Cambridge: Cambridge University Press.

Hughes-Wilson, J. 1999. *Military intelligence blunders.* New York: Carroll and Graf.

Iriye, A. 1999. *Pearl Harbor and the coming of the Pacific War: A brief history with documents and essays.* Boston: Bedford/St. Martin's.

Janis, I. L. 1972. *Victims of Groupthink: Psychological studies of policy decisions and fiascoes.* Boston: Houghton Mifflin.

Jervis, R. 1976. *Perception and misperception in international politics.* Princeton, NJ: Princeton University Press.

Jervis, R. 1983. Deterrence and perception. *International Security* 7: 3–30.

Jervis, R. 2003. The confrontation between Iraq and U.S.: Implications for the theory and practice of deterrence. *European Journal of International Relations* 9: 315–337.

Johnson, D. D. P. 2004. *Overconfidence and war: The havoc and glory of positive illusions.* Cambridge, MA: Harvard University Press.

Johnson, D. D. P., and D. R. Tierney. 2004 Essence of victory: Winning and losing international crises. *Security Studies* 13: 350–381.

Johnson, D. D. P., and D. R. Tierney. 2006. *Failing to win: Perceptions of victory and defeat in international politics.* Cambridge, MA: Harvard University Press.

Kahn, D. 1999. Pearl Harbor as an intelligence failure. In *Pearl Harbor and the coming of the Pacific War: A brief history with documents and essays,* ed. A. Iriye 158–169. Boston: Bedford/St. Martin's.

Kahneman, D., P. Slovic, and A. Tversky. 1982. *Judgment under uncertainty: Heuristics and biases.* Cambridge: Cambridge University Press.

Kaiser, D. 2000. *American tragedy: Kennedy, Johnson, and the origins of the Vietnam.* Cambrige: Harvard University Press.

Klar, Y., D. Zakay, and K. Sharvit. 2002. "If I don't get blown up . . . ": Realism in face of terrorism in an Israeli nationwide sample. *Risk, Decision, and Policy* 7: 203–219.

Kovacs, A. 1997. The nonuse of intelligence. *International Journal of Intelligence and Counter Intelligence* 10: 383–417.

Kuhn, T. S. 1970. *The structure of scientific revolutions.* Chicago: Chicago University Press.

Lashinsky, A. 2006. Chaos by design: The inside story of disorder, disarray, and uncertainty at Google. *Fortune Magazine,* Vol. 154, No. 7, October 2, 2006.

Lebow, R. N. 1981. *Between peace and war: The nature of international crisis.* Baltimore: John Hopkins.

Leigh, A., J. Wolfers, and E. Zitzewitz. 2003. What do financial markets think of war in Iraq? NBER Working Paper 9587. National Bureau of Economic Research, Cambridge, MA.

Levite, A., B. W. Jentleson, and L. Berman. 1992. *Foreign military intervention: The dynamics of protracted conflict.* New York: Columbia University Press.

Levy, J. S. 1994. Learning and foreign policy: Sweeping a conceptual minefield. *International Organization* 48(2): 179–312.

McDermott, R. 2004. *Political psychology in international relations.* Ann Arbor: University of Michigan Press.

Metz, S., and R. Millen. 2004. *Insurgency and counterinsurgency in the twenty-first century: Reconceptualizing threat and response.* Carlisle, PA: U.S. Army War College, Strategic Studies Institute.

Mueller, J. 2005. Simplicity and spook: Terrorism and the dynamics of threat exaggeration. *International Studies Perspectives* 6: 208–234.

Nagl, J. A. 2002. *Learning to eat soup with a knife: Counterinsurgency lessons from Malaya and Vietnam.* Chicago: Chicago University Press.

New Scientist. 2006. Editorial: Kyota in crisis. July 8, 2006, 3.

Perlmutter, A. 1978. Military incompetence and failure: A historical comparative and analytical evaluation. *Journal of Strategic Studies* 1: 121–138.

Perrow, C. 1999. *Normal accidents: Living with high-risk technologies.* Princeton, NJ: Princeton University Press.

Perry, J. M. 1996. *Arrogant armies: Great military disasters and the generals behind them.* New York: John Wiley and Sons.

Petroski, H. 2006. *Success through failure: The paradox of design.* Princeton, NJ: Princeton University Press.

Posen, B. 1984. *The sources of military doctrine: France, Britain, and Germany between the world wars.* Cornell studies in security affairs. Ithaca, NY: Cornell University Press.

Powell, C. 1995. *A soldier's way.* London: Hutchinson.

Rabinovich, A. 2004. *The Yom Kippur War: The epic encounter that transformed the Middle East.* New York: Schocken Books.

Regan, G. 1993. *Snafu: Great American military disasters.* New York: Avon.

Reiter, D. 1996. *Crucible of beliefs: Learning, alliances, and world wars.* Ithaca, NY: Cornell University Press.

Rosenau, W. 2007. U.S. counterterrorism policy. In *How states fight terrorism: Policy dynamics in the West,* ed. D. Zimmermann and A. Wenger, 133–154. Boulder, CO: Lynne Rienner Publishers.

Schacter, D. L., ed. 1995. *Memory distortion: How minds, brains, and societies reconstruct the past.* Cambridge, MA: Harvard University Press.

Schivelbusch, W. 2004. *The culture of defeat: On national trauma, mourning, and recovery.* New York: Picador.

Sears, D. O., L. Huddy, and R. Jervis. 2003. *Oxford handbook of political psychology.* Oxford: Oxford University Press.

Simon, S., and D. Benjamin. 2000. America and the new terrorism. *Survival* 42: 59–75.

Simonsohn, U., N. Karlsson, G. F. Loewenstein, and D. Ariely. 2006. The tree of experience in the forest of information: Overweighing experienced relative to observed information. SSRN Working Paper. Social Science Research Network. October 2006. http//ssrn.com/abstract=521942.

Snyder, J. 1984. *The ideology of the offensive: Military decision making and the disasters of 1914*. Ithaca, NY: Cornell University Press.

Stoessinger, J. G. 1998. *Why nations go to war*. New York: St. Martin's.

Surowiecki, J. 2004. *The wisdom of crowds: Why the many are smarter than the few and how collective wisdom shapes business, economies, societies and nations*. New York: Doubleday.

Tetlock, P. E. 1998. Social psychology and world politics. In *Handbook of Social Psychology*, ed. D. Gilbert, S. Fiske, and G. Lindzey, 868–912. New York: McGraw Hill.

True, J. L. 2002. The changing focus of national security policy. In *Policy Dynamics*, ed. F. R. Baumgartner and B. D. Jones, 155–187. Chicago: University of Chicago Press.

Tuchman, B. 1984. *The march of folly: From Troy to Vietnam*. New York: Alfred A. Knopf.

U.S. Department of Defense. 1971. *The Pentagon papers: United States–Vietnam Relations, 1945–1967*. Washington, DC: U.S. Government Printing Office.

Van Evera, S. 1998. Hypotheses on nationalism and war. *International Security* 18: 5–39.

Van Evera, S. 2003. Why states believe foolish ideas: Non-self-evaluation by states and societies. In *Perspectives on structural realism*, ed. A. K. Hanami, 163–198. New York: Palgrave Macmillan.

Vertzberger, Y. Y. I. 1990. *The world in their minds: Information processing, cognition, and perception in foreign policy decisionmaking*. Stanford, CA: Stanford University Press.

Viola, L., and D. Snidal. 2006. The evolutionary design of international institutions. Working Paper, Program on International Politics, Economics, and Security, University of Chicago, Chicago.

Wirtz, J. J. 1991. *The Tet Offensive: Intelligence failure in war*. Cornell studies in security affairs. Ithaca, NY: Cornell University Press.

Woodward, B. 2005. *State of denial: Bush at war. Part III*. New York: Simon and Schuster.

Chapter 14

# NETWORK ANALYSIS LINKS
# PARTS TO THE WHOLE

FERENC JORDÁN

## The Importance of Being Linked

We live in an increasingly globalized and interconnected world. As the conventional wisdom states, everything is connected to everything else, including humans, ecosystems, and nations. Physicists have examined this complexity, first as a research curiosity, later as a field of science, and currently as a new conceptual paradigm. In this chapter, I present some interesting and notable similarities between ecological and social networks and outline how our knowledge in one field may help in better understanding the other. In particular, I focus on how ecological network theory might help us in thinking about homeland security and understanding modern terrorism in a globalized and interconnected world.

One tool for studying complex systems is network analysis, mostly done in cooperation between physicists, biologists, and scientists belonging to any discipline, having recognized the significance of network studies. Network science is reviving (Strogatz 2001) but not new (e.g., Harary 1959). The main message is even older: we have to study the whole in order to better understand the parts, and vice versa. This is the essence of holistic thinking. What is new is that the traditionally soft, subjective, and occasionally rather mystic statements and approaches of holistic scientists are becoming harder, quantitative, and more scientific. Thanks mostly to network theory and the capacity of computers, now we can calculate to what extent elements are connected to other elements within a network, we can measure complexity and the quantitative relationship between the parts and the whole. Network theory can be applied to different fields (also in interaction with systems theory and cybernetics), either for better understanding the natural systems surrounding us, or for designing better technological systems. Recently, network analysis has been proposed widely as a new tool for exposing and visualizing otherwise probably hidden patterns in political

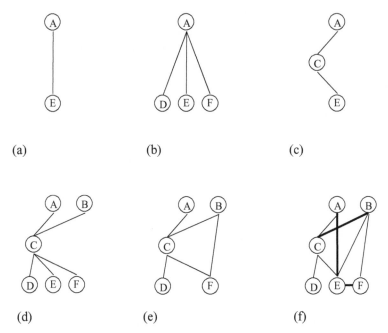

FIGURE 14.1 Six small graphs presenting typical parts of food webs, where nodes and links represent species and their ecological interactions, respectively. Species A is a specialist consumer in some cases (a) but a generalist in others (b). c. An indirect interaction, called "trophic cascade" is shown, where A eats C eating E, and the outcome is a positive effect from A to E. d. A "wasp-waist" food web architecture with several species at low and high trophic levels but only one in the middle (C). e. A loop can be seen containing B, C, and F. f. A food web completed by thick nontrophic interactions (for example, E can be algae living on the surface of the fish A). All these cases are discussed in the text.

sciences (Johnson and Krempel 2004; Keefe 2006; but see also a skeptical critique: Farley 2006). Our network-related knowledge may provide a robust background for defense and homeland security in the future.

Being linked within a network offers both possibilities and constraints. An example from ecology is that the fate of populations of different species depends on their interactions. Specialist predators (like species A in Fig. 14.1a) may go extinct as prey (species E in Fig. 14.1a) go extinct. If a predator feeds on several prey species (species A in Fig. 14.1b), the extinction of one of the prey (e.g., species D in Fig. 14.1b) will not cause its extinction, since it can switch to relying more on alternative resources (species E and F in Fig. 14.1b). A higher number of feeding interactions may increase the resilience of the predator against loosing one of the prey species. However, it can also happen that a particular species goes extinct because of its interactions: kelp forests may disappear if the sea urchin population grows, following the local extinction of sea otters (Estes et al. 1998). Sea otters, feeding on kelp-consuming sea urchins, drive a trophic cascade effect helping the survival of the kelp forest. If the first species dies, the third one may follow it because of their indirect network connection (see Fig. 14.1c, where sea otter is A, sea urchin is C, and kelp forest is E). This situation is familiar in conventional wisdom: the enemies of our enemies tend to be our friends. This is typical in ecology and works also within a group of schoolmates (which I have been studying in Hungary) and frequently in the context of international relations. Thus, to be strongly linked may have both advantages and disadvantages.

Comparisons of different systems frequently reveal that they are surprisingly similar to each other (like the protein networks within our cells and our social networks [see Csermely 2005]). The main focus of current network science is to discover these universal unifying principles (see, e.g., McMahon et al. 2001). Nevertheless, important differences should not be underemphasized: certain network properties, however frequently observed, should not be expected in certain kinds of networks. For example, the scale-free link distribution (a few nodes with many neighbors and a large number of nodes with only a couple of neighbors) found in many kinds of networks applies well to systems like the Internet, but food webs or traffic networks are seemingly constrained not to show this property. Also, some networks do not change while processing information (the Internet or a traffic network remains essentially the same while being used), and the message traveling on the network also remains the same (your email is the same whether you check it in Greenland or Antarctica). Others behave rather like semiconductors (Margalef 1991): information flow and structure influence each other. Ecological and social networks typically belong to the latter kind of networks; this is why it is of particular interest to compare exactly these kinds of networks. Engineers often look for design solutions from nature (e.g., the skipjack nuclear submarine was designed after the height to length ratio of the skipjack tuna). If ecological and social networks seem to share characteristic properties, the question is raised whether we can learn from nature how to design (or, at least, better understand) some of our social networks (including governmental organizations or terrorist networks) in order to stabilize or destabilize them.

## Network Structure: What Is Linked and How?

Most if not all of natural and human-made systems can be represented by graphs where the entities of interest (e.g., people in a social network, airports in a transportation network) are represented by nodes and the focal relationships between them are represented by links between them (see Fig. 14.2 for two examples). The simplest case is just to show whether two

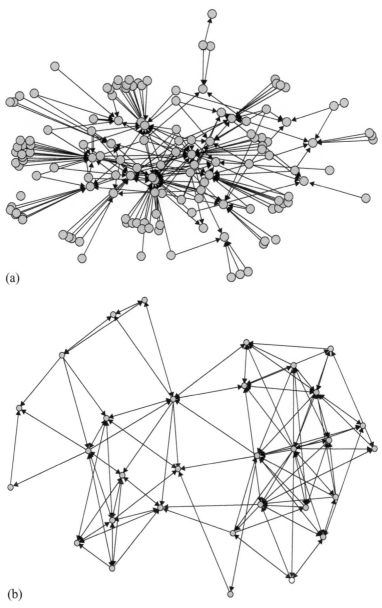

(a)

(b)

FIGURE 14.2 a. An ecological interaction network shows relationships between plants and pollinators in Sweden (nodes are species, and arrows show pollination interactions) (data from Elberling and Olesen 1999). b. A social network shows relationships between children (aged 12) in a Budapest classroom (nodes are individual students, and arrows show strong positive interactions).

chosen nodes are related to each other (being linked or not). This is the topology of a system, providing the most basic information about it. The topology alone gives us some insight on how parts form the whole. However, a graph can provide even more information if the links are weighted (by numbers reflecting some aspect of their quality), signed by the effect one node has on another (positive or negative), or directed (only from node A to node B, only from node B to node A, or in both directions). Adding this kind of information helps to make network analysis more realistic.

Once a system of interest (a human community, an ecosystem, or a traffic network) is represented appropriately by a network, several key parameters can be extracted by techniques of network analysis. These include quantified indirect relationships between two nodes of interest, the centrality and structural importance of a given node (Harary 1961a), the identification of interesting sets of nodes, and global measures of network topology (i.e., link distribution). In general, there is a continuum from local indices (characterizing the most local environment of a node, e.g., the number of its neighbors) to global network statistics (characterizing the whole network, e.g., link density), with intermediate indices in between (considering neighbors of neighbors but dampening with distance, like many centrality indices).

## Local and Global Properties of Networks

Based on the local interaction pattern of a given node, it can be a "star" with a high number of neighbors, it may have a single neighbor in the network, or it can be a "broker" having only a few but highly important neighbors. A whole network can be characterized by its link distribution (like scale-free distribution) and special topologies: it can be a hierarchy, it can contain modules (loosely connected dense groups of nodes), or it can be reliably designed (see box, "Reliable Networks").

Link density (called connectance in ecology) is a global network property correlating with the stability of the network. The early view of ecologists was that sparse networks are less stable, and complexity begets stability; according to an old definition, ecosystem stability is the function of the number of pathways matter can follow in the system while flowing from plants to predators (MacArthur 1955). Later, the contrary was the prevailing opinion; mathematical ecologists demonstrated that complexity has basically destabilizing effects on ecosystems (May 1973). The question is open and needs to be focused, since hundreds of complexity concepts and dozens of stability definitions do exist. Network dynamics can be studied either by simulation models or by descriptive methods. (Unfortunately, in the case of many kinds of networks, there is a massive lack of quality time-series data that track networks through time.)

---

### Reliable Networks

The reliability of a network can be understood as the probability that it will remain connected and functioning in spite of losing parts. To make a network more reliable is very simple if sources are costless: more nodes and more links generally increase stability and reliability. However, if network construction is costly or limited for some reason, network design is an optimization problem. This is one of the key problems of reliability engineering (Aggarwal 1993) and its mathematical background, reliability theory (Barlow and Proschan 1965). The principles of reliable system design seem to have been followed during the evolution of animal development (Molnár 1994), insect societies (Oster and Wilson 1978), and ecosystems (Jordán and Molnár 1999). Reliability, stability, and vulnerability are strongly coupled properties—all influenced by network structure.

---

## On the Strength of Links

An interesting property of a weighted network is the distribution and pattern of link strengths. In ecological networks, the most typically found distribution is that the vast majority of links are relatively weak, and there is a small number of outstandingly strong links (i.e., a predator feeds on one or two strongly preferred and many alternative prey species). Certain models suggest that weak links play surprisingly important roles (McCann et al. 1998) and that the patterning of strong links within the network also matters a lot for stability (long loops should not contain strong links [de Ruiter et al. 1995]). In molecular and social networks the role of weak ties is also a matter of interest: heat shock proteins in cells form many weak molecular links, becoming very important under certain conditions (Csermely 2005; Kovács et al. 2005). It has been suggested that rather than the best friends, the weak personal relationships are the ones providing new kind of information when, for example, looking for a new job. Any careful analysis of link strengths must be understood dynamically, since weight pattern is probably more variable than the topology of links.

## Network Transformation: How Do Networks Change?

Networks do change in time. Nodes can enter and exit a network (e.g., a new student can come to a classroom, or a species may go extinct in an ecosystem), and links can be reoriented or change in quality (weight or sign; see box, "Signed Networks").

> ## Signed Networks
> ___
>
> The concept of structural balance in sociometry says that the sign structure of loops (e.g., A → B → C → A) in networks influences their stability (Roberts 1976). For example, if A affects B positively, B affects C negatively, and C affects A positively, the sign of the loop will be minus. A network is structurally balanced if every loop is positive. A pioneering study by Harary (1961b) used loop signs to better understand international conflicts; the overall signs of relationships between countries were explained by the sign structure of the country network. The key assumption was that countries shape their relationships such that they let all cycles be positive in the network (reducing tension, a consequence of negative loops). This was also probably the first trial of how predictable such a general approach can be.

Measures of stability and robustness are related to the structure of the interaction network. Also, certain structures are fingerprints of past dynamics. For example, scale-free link distribution may imply that the network was formed through a process called preferential attachment (Barabási and Albert 1999); this means that a new node will be linked with higher probability to an existing node already having more neighbors. Structure also gives insight on future behavior. For example, networks with a scale-free link distribution, again, are supposed to be more vulnerable against directed attacks, but also more robust against random errors (Albert et al. 2000).

### Rewiring Networks

Networks can change if the entities alter their interaction pattern or the community as a whole accepts new members (with new links to existing members). The former includes prey switching in ecology (e.g., Gasparini and Castel 1997), or phenotypic plasticity in development (Agrawal 2001) in biological systems. Certain network structures increase stability but only if the trophic links are plastic enough to maintain the connectivity of pathways, that is, switching dynamics is essential for the correct functioning of a reliable structure (Jordán et al. 2003). In fact, switching ability and antipredator behavior (see Blumstein, this volume) are among the most important population dynamical details that complement network analysis based on simple topology (cf. Johnson and J. Madin, this volume). Changing friends or business partners can also cause plasticity in network structure. Depending on the actual system, there can be high variability in how new members become related to others in the network. A new student entering a classroom becomes instantly part of a very transparent network where he may find many new friends and many new enemies. By contrast, a recruit to a terrorist network may be given only one or two contacts and thus be a large distance from the rest of the network.

Small-scale temporal plasticity can be an important feature of a network: terrorist networks probably change their structure drastically before a coordinated action, and the resulting topology is more suitable for operation (Krebs 2002). It is suggested that loosely connected terrorists form a network where their average distance is surprisingly large (i.e., if A wants to send a message to Z through others, it may need several transmitters). However, before a coordinated action, shortcuts appear in the network (e.g., J establishes a contact to K, not seen before), and the average distance will become surprisingly small (Krebs 2002). At the same time, these last-minute changes do not alter the basic network structure: the network remains redundant with high structural equivalence and turnover of nodes, and a relative lack of centrality (Rothenberg 2002). Since communication and coordinated action are key issues, fragmenting terrorist networks and isolating individuals comprise a candidate defense strategy (Lafferty, Smith, and Madin, this volume), because proper functioning requires dynamical connectedness.

## Adding New Nodes to Networks

Beyond repatterning the links, compositional changes happen if a new species invades an ecosystem. "Invaded Networks" (see box) gives an example for how the success of invasion depends on network structure (on the "health" of the resident community), and in particular, on the position of the invader. "Invasion" in a social network context simply means that a new person enters a group. In political life, new ideas probably spread more easily if the structure of the society is damaged, for example, following a revolution or a war, or in economical crisis (especially if these result in a generally lower education level [Ehrlich and Levin 2005]). Wealth probably results in more conservative and less invadable communities. If we consider terrorist persons themselves as infected agents, we can say that their resistance against infection depends on the healthy structure of the social, religious, and economical network they live in. Becoming a terrorist is probably a highly context dependent event; the network neighborhood may determine it to a large extent.

## Invaded Networks

Ecological communities are continuously invaded by other species. The success of invasion depends on how the new species fits into the preexisting community structure (Sugihara 1984). Coadjusted resident species may have the homefield advantage, or alternatively, the invader restructures the community (Pimm 1991). A famous example for the latter is the invasion of the exotic fish *Cichla ocellaris* in Gatun Lake, Panama (Zaret and Paine 1973), causing the extinction or emigration of many other species from the lake. The success of invasion depends on the "health" of the resident community; the spectacular success story of the jellyfish *Mnemiopsis lleidyi* in the Black Sea was possible partly because of the former overfishing of the marine community (Shiganova 1998). This invasion led to a famous regime shift: the advantageous topological position of the jellyfish in the food web (in wasp-waist position, see Fig. 14.1d) helped it in mostly outcompeting a formerly dominant fish, the Black Sea anchovy, from the system (Shiganova 1998; Jordán 2006). Related questions for social network studies are how and to what extent can people invading a network (e.g., new members of a terrorist organization, or spies infiltrating the organization) change the network function and stability.

## On the Role of Flexibility

Networks differ from the viewpoint of the extent, mechanisms, and the time scale of plasticity. For example, army command hierarchies are more rigid, human friendship networks are definitely more changeable, and prey switching of predators can be very fast but transitory. An important feature from this viewpoint is the way these systems are organized. Governmental administration networks are traditionally more hierarchical than private sector organization networks (e.g., Dekker 2002), and the private sector typically maintains more heterogeneity (Prescott, this volume). As a consequence, governmental networks are less flexible and adaptable (Johnson and E. Madin, this volume). Both the army and the Mafia are hierarchies, unlike terrorist networks, that also have higher-order cores but primarily are composed of loosely connected, semiautonomous units (note also that the hierarchy within terrorist networks is strongly obeyed in some times and not in others [Thayer, this volume]). Centrally controlled, hierarchically controlled, externally controlled, and democratically self-organizing systems differ in the reversibility and speed of change (Rischard 2002). For example, an externally controlled system (like a fish school driven by the presence of a predator, or a marmot group hibernating in winter) must follow the changes of its environment, while certain hierarchies can change quickly (e.g., male deer fighting for females). In the first case, the interaction structure is important but external factors can overrule it, while in the second, the interaction network is of primary importance.

Ecosystems can be regarded as complex adaptive systems (e.g., Levin 1999; Levin et al. 2001) and are able to change rapidly if needed, because of both self-organization and selection at lower levels; small, accumulating changes may induce bottom-up effects leading to complex adaptations, and the exact mechanisms and possibilities depend on network topology. One possible similarity between social and ecological systems is that macroscopic patterns emerge mostly from accumulating microscopic (lower-level) changes (Levin 1999, 104), since typically there is no selection at the system level. Moreover, it is suggested that coexistence of agents in political systems also depends on their interaction structure ("political ecosystem," [Prescott, this volume]).

Large changes in network structure may also be consequences of threshold effects. Regime shifts of ecosystems are examples of sudden changes in the interaction structure, driven by either a large change in the abiotic environment (external control) or a positive feedback loop triggered by a threshold effect (see box, "Invaded Networks," for a regime-shift example, as well as work by Scheffer and Carpenter [2003]). The role of positive feedback loops has been frequently emphasized before (e.g., Ulanowicz 1989) (a feedback loop is illustrated in Fig. 14.1e, where C feeds on F, B feeds on C, and F is influenced somehow by B). In this sense, we should expect less major shifts appearing in loopless hierarchies (cf. the punctuated change and relative rigidity of many domestic intelligence units [Johnson and E. Madin, this volume]), which has the beneficial effect of conferring stability, but the negative effect of making the organization resistant to adaptation.

## The Role of Space: Limits and Possibilities

Both the structure and the dynamics of kinds of networks are constrained by system-specific factors (Amaral et al. 2000), possibly leading to deviations from universal governing principles. For example, airways networks can grow by preferential attachment but road networks cannot. Global terrorism, strongly depending on (mostly virtual) information flow (cf. Atran, this volume; Atran cited in Ariza 2006), may exist only in a world where topography matters less than the topology of interactions. Our societies seem to be in transition; topography is still responsible for roughly two-thirds of friendship tie patterns (Liben-Nowell et al. 2005).

Spatial constraints, distances, and boundaries result in defining a patchwork of social groups. Preserving group identity may be more difficult in a spatially unconstrained and unstructured world. One function of global networklike organizations, such as terrorist networks, can be to establish niches of high local density or interaction within a regional or global network. This can be highly important in an overhomogenized world. For some people, joining a terrorist network may be helpful in finding group identity. A further reaction to globalization is "glocalization": building a virtual, global neighborhood (networked individualism [Wellman 2002]). These new subnetworks are not "organically" developed interaction webs—just reactions to abnormally fast processes (a kind of norm change [Ehrlich and Levin 2005]). From this viewpoint, terrorism is partly a negative consequence of globalization (Ariza 2006).

In several kinds of networks, links between any two nodes are possible, while certain networks face serious structural constraints. The most typical case is topographical constraints: networks realized in space (e.g., on a two-dimensional surface) cannot have any topology. For example, it is nonsense to build a highway directly from Los Angeles to New York City with no exits to other cities. In the route map of the United States, Los Angeles and New York City cannot be directly linked. In contrast, in a social network of human beings, basically everyone can be a friend of everyone else: there is no (spatial) constraint on topology. In our increasingly interconnected and globalized world, a number of such topographical constraints are relaxed, with important consequences. In nature, spatial heterogeneity is frequently the key to coexistence. Survival of weak competitors may be possible only in temporary refuges; their lack is one of the most serious problems of global overfishing in marine ecosystems. Following a long debate in the 1970s, we can probably accept the conclusion that compartments in food webs appear only if they represent spatially distinct habitats, and interactions within a given habitat are typically not compartmentalized (Pimm 1982). For many rare species, the alteration of local extinction and recolonization is a typical phenomenon, underlying the importance of space in dynamical processes.

In both nature and society it seems to be true that cooperation is more stable if feedbacks are strong and the interactive partners are within a smaller distance in the social network (Ehrlich and Levin 2005); we tend to cooperate if positive effects feed back quickly. It could be an important factor behind terrorism that the negative effects are dispersed in a nonspatial, global social network instead of quickly feeding back to the agents. In this sense, spatial constraints can result in higher responsibility (are not people more friendly in small towns?).

## Key Nodes in Networks: Key to Understanding?

In both characterizing the structure and understanding the dynamics of networks, we have to consider their heterogeneous nature. This is meant here in two senses. First, the nodes and links are of different quality.

### Subway Networks

Targets of terrorists are typically not easy to predict but could be highly useful for optimizing defense strategies. A simple network analysis revealed that the three London Underground stations attacked on July 7, 2005 might have been chosen based on network analysis (Fig. 14.3). Widely available software helps in calculating the positional importance of stations in maintaining the connectivity of the subway network (based on betweenness centrality). Based on a couple of simple and plausible assumptions, it was found that the combination of the three stations that were bombed ranked second best (in terms of destructive capability) out of more than 3.2 million possible combinations (i.e., picking 3 stations out of 270) (Jordán, in press). Note that in another case (sarin gas attack in Tokyo), ranks of target stations were much lower, but this is not surprising because in that case people were the target, not infrastructure. Thus, the stations attacked were not of high structural importance but were very crowded.

Second, even if they are of the same quality, their topological positions are different; they are of different centrality. Heterogeneous networks represent a kind of organization between strict hierarchies and homogenous networks (Margalef 1991). One possible solution for understanding complex systems is to focus on wisely selected members. In a biological network, these are keystone species (in food webs) or hub metabolites (in molecular networks). Key nodes of a social network can be the key players of a human group or the key agents in a terrorist network (the aces of the 52 cards).

The local network neighborhood of nodes is called their social field in social networks or trophic field in food webs (these can be quantified by centrality indices). The interactive fields of key nodes are of the highest interest. Some studies demonstrate the relationship between local network position and extinction probability in model food webs (Jordán et al. 2002; see also box, "Subway Networks," for a technological example of critical nodes in networks). Understanding human group dynamics can be easier if the sociometric star person is observed, and understanding decision making is easier if focusing on key policy makers with high responsibility and power (Prescott, this volume). Focusing on well-chosen members of a network is also practical, considering the huge mass of information in certain systems; even the most modern computational capacity is limited, so information overload should be reduced if possible (Keefe 2006). Still, it is

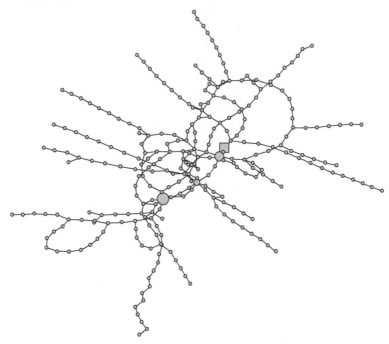

FIGURE 14.3 The London Underground network. Nodes and lines are stations and tunnels between them, respectively. The three larger circles are the three structurally most important stations out of 270; radius shows their importance rank (based on betweenness centrality). The July 7, 2005 attacks targeted the first-ranked one (largest circle: Liverpool Street), the third one (King's Cross), and the one represented by the square (Edgware Road) instead of the second one (second-largest circle: Baker Street), being next to it. This choice is highly improbable by chance. The possibility is raised that some kind of network analysis had been made when preparing these attacks.

important to consider global network properties as well as the underlying factors that drive social networks (group identity [see Villarreal, this volume]).

In the social biology of animals, interest is increasing in how social structure influences group dynamics and determines the individuals acting as key players (dolphins: Lusseau 2003; primates: Flack et al. 2005). In case of social primates, knocking out particularly important individuals ("effective policy managers") causes fragmentation of the social network (Flack et al. 2005). Recent studies report on quantifying the positional importance of both politicians (Johnson and Krempel 2004) and terrorist leaders (Krebs 2002) in their social networks.

## Conclusions: Implications for Policy

As in many fields of science and technology, we can always learn from nature (e.g., Oster and Wilson 1978; Cohen 1988). We can investigate which natural networks function well, which ones do not and why, which ones are plastic enough to change rapidly, and which ones are robust enough against failures. Typically we are not sure which results are to be applied in our society, even if we have a menu of successful case studies in nature. The three basic strategies are (1) to optimally design, (2) to understand and try to control, or (3) to predict and utilize networks, all wherever feasible. The first strategy is applied in designing power grid networks, the second is a possible background of public health programs (Borgatti 2003), and the third applies to network-based knowledge in stock analysis.

For security considerations, we may hope to identify natural conditions that will cause a network to collapse, obviating the need for explicit action. The "eight commandments of environmental management" (Levin 1999) present a number of general recommendations, derived from the analysis of biological systems, that might be considered when understanding, predicting, and managing the behavior of complex systems (see also Vermeij, this volume). Some of these are about networks (tightened feedback loops, maintained redundancy and heterogeneity, modular structure) and also seem to be of possible relevance in better understanding social networks, both defense and terrorist types.

Networks ranging from transportation to organization to communication systems and international relations could be better organized for security purposes. Using insights from network theory, we can design resilient or resistant systems, the former managing, the latter escaping failures. One key to good design can be an optimal level of flexibility (partly determined by the nature of the system). Both hierarchies and democracies are disadvantageous under certain conditions: a hierarchy can be less sensitive to the actual environment (because of its limited number of decision makers), while a democratic network can produce slower decisions (because it requires more communication). Probably the most adaptable natural networks are somewhere in between, with local hierarchies in a large-scale democratic network. Rischard (2002) suggests that a solution for global policy-making problems could be a "networked global governance," with minimalized hierarchy and fast delivery time. This structure is better for catalyzing communications than for transmitting decisions.

One consideration of interest is to try to foster the "organic" development of social and ecological systems. Sudden changes often lead to failure and collapse. Overfishing, drastically changing interaction patterns, opens plenty of possibilities for invasive species, while wars and economical crises, drastically altering social interactions, make strange and dangerous ideas spread easily. For a concrete example, there was a lot of criticism after Hurricane Katrina destroyed New Orleans about the mistake of putting the Federal Emergency Management Agency into the newly and artificially formed Department of Homeland Security. In this case, a sudden, large structural change disturbed the behavior of the system.

Since there are many kinds of interactions between the nodes in both social and ecological communities, it seems to be helpful to represent them by networks with many kinds of links. In such a representation, some links may represent positive effects, whereas others represent spreading dangers. The desired goal is to try to shape networks where the former kinds of links are densely connected and well maintained, while the latter kinds of links are sparsely connected and cut efficiently. The same links can even serve both positive and negative purposes; it has already been suggested that the Internet, being highly important for the communication within terrorist groups, could also be used as a tool for spreading alternative messages (Thayer, this volume; Atran, this volume) or false information (Ariza 2006). Note that infrastructure quality also has always contrasting effects, either helping or limiting both the defense and the insurgent sides (Johnson and J. Madin, this volume). In conservation practice, helping one species has negative effects on another; conflicts between species protection programs are persistent. These conflicts immediately rise as soon as we try to protect "whole" ecosystems instead of selected species (cf. the protection of large, highly productive systems in order to reduce vulnerability [Vermeij 2004]). One approach to this problem is to also incorporate nontrophic links in food webs (cf. Fig. 14.1f, where thin links represent feeding relationships and thick links represent nontrophic effects, such as facilitation or mutualism). Similarly, a social network may represent various kinds of interactions simultaneously (e.g., friendship, business relationship, common educational background), and social network analysis could provide a tool for studying these kinds of links in parallel.

The ultimate answer must be that prevention is the key to success. Decreasing the negative, homogenizing side of globalization, maintaining spatial heterogeneity, reorganizing existing social networks less rapidly (i.e., favoring slower democratic processes), or building more flexibly functioning defense networks are all tools for achieving a higher level of security. It is also important to collect as much reliable data on network time series as possible, for comparative purposes. It may be possible that our global, increasingly democratic and homogeneous social network is a sine qua non condition for modern terrorism (Rothenberg 2002). If having risen, these networks can be studied and fought, but preventing their rise might be even a more efficient solution. Finally, we must understand what the most characteristic patterns on network dynamics are in nature and in our society, to what extent they are predictable, and how to utilize rather than fear the constantly changing networks around us.

## Future Research and Open Questions

*Future research* should be focused on improving the techniques of descriptive network analysis, in a multidisciplinary and comparative way. Instead of studying the global statistics of static networks, we have to describe and compare the behavior of key network elements as determinants of group dynamics. We have to test which network parameters are the best predictors of group dynamics. We have to collect more data on terrorist networks and identify their most relevant structural and dynamical properties. On the other side, we may revise the structure of networks in administration and security.

The major *open questions* are (1) to what extent terrorist networks are similar to typical social networks, in general, (2) whether observed structural similarities between natural and social systems may be explained by similar mechanisms, (3) to what extent will we be able to reliably predict the behavior of complex ecological and social systems, (4) whether it is possible to prevent or reduce terrorist activity by using our knowledge about social networks, and more generally, (5) whether and how we can more effectively utilize the advantages of globalization instead of tolerating its negative consequences.

### ACKNOWLEDGMENTS

I thank Raphael Sagarin, Geerat Vermeij, Katerine Smith, Péter Csermely, Lajos Rózsa, and András Takács-Sánta for excellent comments on the manuscript. I am especially grateful to Dominic Johnson and Vera Vasas for excellent discussions on networks, and the editors for the invitation to the very interesting NCEAS working group meetings in Santa Barbara, California. My research is fully supported by Society in Science: The Branco Weiss Fellowship, ETH Zürich, Switzerland.

### GLOSSARY

*Betweenness centrality.* One of the centrality indices characterizing the positional importance of nodes within a graph. A central node is characterized by a high ratio of shortest pathways between every possible pair of other nodes containing the node of interest.

*Broker.* A graph node of high positional importance, still having only a few neighbors. Its neighbors are of very high importance.

*Centrality.* A quantifiable property of nodes in networks, characterizing their position. High centrality means positional importance (e.g., the middle node in a starlike design). There are many network indices for centrality, reflecting different aspects of positional importance (e.g., betweenness centrality). The centrality of links can also be defined.

*Constraints.* Any physical law or less-universal effect limiting the possibilities for some property of some object. For example, gravity unavoidable influences the embryonic development of frogs; also, food chains cannot be infinitely long because of the loss of transferred energy.

*Directed graph.* A network where each link has a direction (i.e., "from A to B" and "from B to A" are not equivalent). For example, command charts of armies depict who obeys whom.

*Feedback loop.* An indirect effect pathway finally affecting its origin. For example, A affects B, B affects C, and C affects A.

*Food web.* A graph representing the feeding relationships between coexisting species (who eats whom). Species are nodes and feeding interactions are links. A food web is composed of food chains connecting consumers (lion or shark) to producers (grass or phytoplankton). Nodes may be species or functional groups (e.g., benthic feeders).

*Graph.* Mathematical representation showing well-defined relationships (links) between objects (nodes). Here we use this term synonymously with 'network."

*Hierarchy.* A directed graph with no cycles (e.g., A to B, B to C, and C to A).

*Link density.* The number of links in a network, compared to the maximal possible number.

*Module.* A well-defined small region in a network, appearing frequently. For example, A eating B and C plus B eating C is called intraguild predation in food webs and called a feedforward loop in molecular networks. Software helps in searching for modules and statistically evaluating them. Modules are also called motifs.

*Network dynamics.* Changes in the composition of nodes and/or links in a network.

*Nontrophic interaction.* An ecological interaction between two species that is independent of predatory effects (e.g., giving shelter or substrate to another species).

*Preferential attachment.* A network growth model saying that a new node in a network will be connected with higher probability to existing nodes with more neighbors.

*Recolonization.* Colonization of a species to a habitat where it had been previously living but went extinct.

*Redundant.* Objects with (partly) overlapping function. For example, if several pollinators coexist with a plant, the extinction of one may not be disastrous for the plant, since the others still can maintain pollination.

*Scale-free distribution of links.* This characterizes a network where only a couple of nodes have many neighbors and the majority of nodes have very few neighbors.

*Self-organization.* A number of microscopic, simple interactions producing macroscopic, qualitative changes that result in well-recognizable patterns.

*Shortcut.* A new link between two distant nodes of a network.

*Signed graph.* A network where each link has a sign (either positive or negative). For example, the social network of a classroom may have "like" and "dislike" links between students.

*Social network.* A graph representing relationships between human individuals. For example, a social network shows who likes whom in a classroom. Social networks may also depict animal groups or human-made systems if defined broadly (e.g., the Internet).

*Sociometry.* A quantitative branch of sociology, applying mathematical tools for analyzing human groups. The key tool of sociometry is social network analysis.

*Structural equivalence.* Two or more graph nodes are structurally equivalent if they link to similar neighbors. For example, two species of zooplankton can be structurally equivalent in a food web if they are eaten by the same fish and feed on the same phytoplankton species.

*Switching (prey switching).* An ecological phenomenon, when a consumer drastically changes its feeding habits from a strongly preferred and exploited prey to another one. Switching changes the importance of links in the network.

*Threshold effect.* Small, accumulating changes leading to a major, qualitative change after reaching a well-defined value.

*Topology.* A branch of mathematics studying neighborhood relationships. In the topology of a human social network the only thing that matters is who is linked to whom. If we consider the quality of either links or nodes, we call it a network structure.

*Topography.* Very simply, topology realized in space. Beyond neighborhood relationships (what is linked to what) it is also about metrics (e.g., distance in miles).

*Weighted graph.* A network where each link has a number quantifying its quality. For example, a link in a food web can be quantified by the value of energy transfer from prey to predator.

## REFERENCES

Aggarwal, K. K. 1993. *Reliability engineering.* Dordrecht: Kluwer Academic Publishers.

Agrawal, A. A. 2001. Phenotypic plasticity in the interactions and evolution of species. *Science* 294: 321–326.

Albert, R., H. Jeong, and A.-L. Barabási. 2000. Error and attack tolerance of complex networks. *Nature* 406: 378–381.

Amaral, L. A. N., A. Scala, E. Barthélémy, and H. E. Stanley. 2000. Classes of small-world networks. *Proceedings of the National Academy of Sciences USA* 97: 11149–11152.

Ariza, L. M. 2006. Virtual jihad. *Scientific American,* January, 18–20.

Barabási, A.-L., and R. Albert. 1999. Emergence of scaling in random networks. *Science* 286: 509–512.

Barlow, R. E., and F. Proschan. 1965. *Mathematical theory of reliability.* New York: Wiley and Sons.

Borgatti, S. P. 2006. Identifying sets of key players in a social network. *Computational and Mathematical Organization Theory* 12: 21–34.

Cohen, J. E. 1988. The counterintuitive in conflict and cooperation. *American Scientist* 76: 576–584.

Csermely, P. 2005. *Weak links: Stabilizers of complex systems from proteins to social networks.* Berlin: Springer Verlag.

Dekker, A. 2002. Applying social network analysis concepts to military C4ISR architectures. *Connections* 24 (3): 93–103.

De Ruiter, P. C., A.-M. Neutal, and J. C. Moore. 1995. Energetics, patterns of interaction strengths, and stability in real ecosystems. *Science* 269: 1257–1260.

Ehrlich, P. R., and S. A. Levin. 2005. The evolution of norms. *PLoS Biol* 3 (6): e194.

Elberling, H., and J. M. Olesen. 1999. The structure of a high latitude plant-flower visitor system: The dominance of flies. *Ecography* 22: 314–323.

Estes, J. A., M. T. Tinker, T. M. Williams, and D. F. Doak. 1998. Killer whale predation on sea otters linking oceanic and nearshore ecosystems. *Science* 282: 473–476.

Farley, J. D. 2006. The N. S. A's math problem. *New York Times,* May 16.

Flack, J. C., F. B. M. de Waal, and D. C. Krakauer. 2005. Social structure, robustness, and policing cost in a cognitively sophisticated species. *American Naturalist* 165 (5): E126–E139.

Gasparini, S., and J. Castel. 1997. Autotrophic and heterotrophic nanoplankton in the diet of the estuarine copepods *Eurytemora affinis* and *Acartia bifilosa. Journal of Plankton Research* 19: 877–890.

Harary, F. 1959. Status and contrastatus. *Sociometry* 22: 23–43.

Harary, F. 1961a. Who eats whom? General Systems 6: 41–44.

Harary, F. 1961b. A structural analysis of the situation in the Middle East in 1956. *Journal of Conflict Resolution* 5: 167–178.

Johnson, J. C., and L. Krempel. 2004. Network visualization: The "Bush Team" in Reuters News Ticker 9/11 to 11/15/01. *Journal of Social Structure* 5: 1. Also available at www.cmu.edu/joss/content/articles/volume5/JohnsonKrempel/.

Jordán, F. 2006. Topological constraints on the dynamics of wasp-waist ecosystems. In *Regime shifts in aquatic ecosystems*, ed. V. Velikova and G. Dashkalov. UNESCO-Roste/BAS Press, 58–67.

Jordán, F. In press. Predicting target selection by terrorists: A network analysis of the 2005 London underground attacks. *International Journal of Critical Infrastructures.*

Jordán, F., and I. Molnár. 1999. Reliable flows and preferred patterns in food webs. *Evolutionary Ecology Research* 1: 591–609.

Jordán, F., I. Scheuring, and G. Vida. 2002. Species positions and extinction dynamics in simple food webs. *Journal of Theoretical Biology* 215: 441–448.

Jordán, F., I. Scheuring, and I. Molnár. 2003. Persistence and flow reliability in simple food webs. *Ecological Modelling* 161: 117–124.

Keefe, P. R. 2006. Can network theory thwart terrorists? *New York Times*, March 12.

Kovács, I. A., M. S. Szalay, and P. Csermely. 2005. Water and molecular chaperones act as weak links of protein folding networks: Energy landscape and punctuated equilibrium changes point towards a game theory of proteins. *FEBS Letters* 579: 2254–2260.

Krebs, V. E. 2002. Mapping networks of terrorist cells. *Connections* 24 (3): 43–52.

Levin, S. A. 1999. *Fragile dominion*. Cambridge, MA: Perseus Publishing.

Levin, S. A., J. Dushoff, and J. E. Keymer. 2001. Community assembly and the emergence of ecosystem pattern. *Scientia Marina* 65 (Suppl. 2): 171–179.

Liben-Nowell, D., J. Novak, R. Kumar, P. Raghavan, and A. Tomkins. 2005. Geographic routing in social networks. *Proceedings of the National Academy of Sciences USA* 102: 11623–11628.

Lusseau, D. 2003. The emergent properties of a dolphin social network. *Proceedings of the Royal Society of London B* (Suppl.) 270: S186–S188.

MacArthur, R. 1955. Fluctuations of animal populations, and a measure of community stability. *Ecology* 36: 533–536.

Margalef R. 1991. Networks in ecology. In *Theoretical studies of ecosystems: The network perspective*, ed. M. Higashi and T. P. Burns, 41–57. Cambridge: Cambridge University Press.

May, R. M. 1973. *Stability and complexity in model ecosystems*. Princeton, NJ: Princeton University Press.

McCann, K., A. Hastings, and G. R. Huxel. 1998. Weak trophic interactions and the balance of nature. *Nature* 395: 794–798.

McMahon, S. M., K. H. Miller, and J. Drake. 2001. Networking tips for social scientists and ecologists. *Science* 293: 1604–1605.

Molnár, I. 1994. Developmental reliability and evolution. In *Interplay of genetic and physical processes in the development of biological form*, ed. D. Beysens, G. Forgács, and F. Gaill, 161–167. Singapore: World Scientific.

Oster, G. F., and E. O. Wilson. 1978. *Caste and ecology in the social insects*. Princeton, NJ: Princeton University Press.

Pimm, S. L. 1982. *Food webs*. Chapman and Hall.

Pimm, S. L. 1991. *The balance of nature?* Chicago: University of Chicago Press.

Rischard, J. F. 2002. Global issues networks: Desperate times deserve innovative measures. *Washington Quarterly* 26 (1): 17–33.

Roberts, F. S. 1976. *Discrete mathematical models.* Philadelphia: Prentice Hall.

Rothenberg, R. 2002. From whole cloth: Making up the terrorist network. *Connections* 24 (3): 36–42.

Scheffer M., and S. R. Carpenter. 2003. Catastrophic regime shifts in ecosystems: Linking theory to observation. *Trends in Ecology and Evolution* 18: 648–656.

Shiganova T. A. 1998. Invasion of the Black Sea by the ctenophore *Mnemiopsis leidyi* and recent changes in pelagic community structure. *Fisheries Oceanography* 7: 305–310.

Strogatz, S. H. 2001. Exploring complex networks. *Nature* 410: 268–276.

Sugihara, G. 1984. Graph theory, homology and food webs. *Proceedings of the Symposium on Applied Mathematics* 30: 83–101.

Ulanowicz, R. E. 1989. A phenomenology of evolving networks. *Systems Research* 6: 209–217.

Vermeij, G. 2004. Ecological avalanches and the two kinds of extinction. *Evolutionary Ecology Research* 6: 315–337.

Wellman, B. 2002. The rise (and possible fall) of networked individualism. *Connections* 24 (3): 30–32.

Zaret, T. M., and R. T. Paine. 1973. Species introduction in a tropical lake. *Science* 182: 449–455.

# Chapter 15

# A HOLISTIC VIEW OF NATURAL SECURITY

RAPHAEL D. SAGARIN

## The Tension

When considering the role of ecology and evolution in a discussion of contemporary societal issues, there is a fundamental tension between two opposing forces. First is the fact that we as individuals and societies are intimately shaped by our natural history. Second is the reality that our daily lives, our technologies, and our societal institutions appear to operate independently of evolutionary control and thus would not, or could not, benefit from an evolutionary treatment. Chapters in this volume have examined components from each side of this dichotomy. Here I argue that when taken as a whole, a natural history approach to security can reconcile the tension between strong evolutionary control and evolutionary independence of societal actions. I briefly discuss the justification for these two viewpoints, through examples from this book and other research, and demonstrate how they fit into four operational concepts of natural security. What emerges is that both an awareness of how natural forces control the security environment and an understanding of where society operates freely of evolutionary and ecological control are key components to more efficient and effective security analysis and planning.

Evolutionary and ecological dynamics operate at a range of time scales to intimately shape our current security environment. Most directly, our body form and physiology set limits on the environments in which we live, and the stressors and attacks we can withstand. Indirectly, outward characteristics associated with different races arose as evolutionary responses to different ancestral environments but now play a role (how strong a role is an ongoing matter of debate) in tensions between populations and in the relative vulnerabilities of different populations. In turn, racial identity and its effects are related to evolutionarily derived behaviors. Indeed many behaviors that affect the security environment arose from selection that

261

occurred over several hundred thousand generations living as hunter-gatherers. If they do not impose a significant cost relative to their benefits, the behaviors are not likely to be erased by less than 10 generations living in an industrial society.

Outside our human bodies, our security environment has also been shaped by billions of years of biogeochemical processes: the proliferation and decay of life forms, the mass movement of geological plates, volcanism, and the chemical processes that link the biological and geological components of our planet. Jared Diamond has argued that the distribution of key resources, such as favorable climates, plant and animal species amenable to domestication, as well as the movement of pathogenic organisms, were driving forces in the differential development of human societies (Diamond 1999). Geerat Vermeij has tied biogeochemical processes at a range of time scales to an economic system of resources that shapes (and is shaped by) interactions between species, from bacteria to humans (Vermeij 2004).

Yet despite the many pathways by which natural processes control our security environment, it is easy to overlook the role of evolution in contemporary geopolitical issues. On most days, most humans—especially those making decisions that directly affect societal security—are not engaged in a struggle for survival, which to many is the defining characteristic of Darwinian evolution (that this viewpoint ignores the reality that on a daily basis starvation, disease, and conflict take thousands of human lives only reiterates the tension between evolutionary and ecological control over society and our perceived disconnect from such control). Moreover, at the level of institutions within society, it is easy to be skeptical of the notion that strong evolutionary forces are at work. In a democracy, voters are a potential selective mechanism for generating effective policy makers, but their strength is diffused by political impediments such as (in the United States) gerrymandered districts that predetermine electoral outcomes and committee chairmanships that provide disproportionate power to certain individuals. Likewise, budgets might be seen as the selective force on institutions managed by elected or unelected officials. Yet here again, political quirks, such as congressional "earmarks," in which funds are funneled without debate to pet projects, severely weaken the selective strength of the budget process. Routinely, failing or unnecessary projects (which in a natural evolutionary system would have been selected against) are maintained or even given more funding. In theory, free economic markets contain elements akin to natural selection, and discussion about firms in a free market are often caged in evolutionary terms ("only the fit survive"). In reality, however, legal constraints, subsidies, and incomplete information (for example, oil markets do not capture information about environmental damage or human conflict associated with oil consumption) limit the effectiveness of selective forces acting on market players.

There are elements of both perception and reality in the apparent disconnection of our lives from natural forces. Unlike biology for which Dobzhansky (1973) said, "Nothing . . . makes sense except in the light of evolution," the fields of societal study that have primarily analyzed security, from history to international relations, have largely operated without reference to evolutionary theory (Gregory Dietl's chapter here provides a rare exception). This has perhaps reinforced the notion that evolutionary theory is unnecessary to assess problems in society. Even to the extent that societies actually "evolve," it is in the form of evolution of norms (societal conventions or customs), which is characteristically different than genetic evolution. As an example, norms can be transferred within and between generations in any direction, whereas genes are generally transferred only between generations, from parent to offspring (Ehrlich and Levin 2005).

Yet, whatever the true relative contribution of evolutionary forces to societal outcomes, it is clear that through our cognition, behaviors, and societal institutions, we have found ways to dampen or sidestep the relentless and merciless environmental control faced by most organisms on Earth. On one hand, this is a great detriment. No organic, self-organized force ensures that our systems are adequate for survival. On the other hand, it can be seen as an opportunity. We have created a space in which we can analyze security problems from a detached standpoint, anticipate likely outcomes, and design specific responses based on that information.

This is no more than what security analysts and policy makers are frantically doing today. What is missing, though, is the force of selection, which Gregory Dietl identifies (this volume) as a key universal principle of Darwinian evolution. With an evolutionary perspective we seek to initiate artificial selection by testing security systems, deliberately replicating successful outcomes of those tests, adapting marginal performers to work better, and eliminating failures. Thus, we can create systems that evolve both in response to, and in anticipation of, threats in the environment. And just as no organism evolves independently of its environment, all of our selective tinkering must be done with a fundamental consideration of the evolutionary roots and ecological context of the security situation.

Thus, a natural history of security allows us, or forces us, to reconcile the tension between our evolutionary roots and our modern detachment by taking a holistic view. The application of evolutionary study to security is not merely a matter of copying what we see in nature. Rather, it should be a process of simultaneously considering how we can adopt the most successful features in natural systems while keeping a sharp eye on how our own evolutionary history shapes the security environment today. I address both sides of this tension in the four operational concepts of natural security that I introduced at the beginning of this volume and return to here.

## Four Operational Concepts of Natural Security

### Organization

As Vermeij has stressed in this volume and elsewhere, the most successful model of organization for survival and resilience incorporates semiautonomous units that sense and respond to the environment and can communicate with one another while operating under only limited central control. This model of organization appears repeatedly at a wide range of biological scales. At the cellular level, genetically controlled organelles have specialized functions and can communicate results of these functions within and among cells through chemical channels. The immune system incorporates this model by sending out specialized agents to both detect and destroy pathogens. Those that are successful in their search are upregulated, while those that fail are downregulated to reduce energy use and increase precision of the response. Organisms show this organization in a variety of ways, from clonal aggregations with specialized food gathering, defense, and reproductive clones, to specialized body parts that according to Vermeij, tend to be more prevalent in more recently developed species within a given clade (Vermeij 2004).

This organizational structure is mimicked in higher levels of biological organization. Theoretical work by Simon Levin and colleagues at the scale of ecosystems shows that the autonomous self-motivated actions of individual organisms impart an overall quality of resilience to the ecosystem (Levin 1999). Sapolsky (2006) notes that human hunter-gatherers form malleable networks made of tight local interactions and more diffuse interactions over larger spaces. Network science, as described by Barabasi (2003) and others, shows promise as a tool to describe many of these structures and bridge the cultural divide between social and biological sciences, given that protein networks, social relationships, ecosystems, and global transport systems can be described with strikingly similar maps. As Ferenc Jordán recognizes (this volume), however, the real value of networks will come by studying their dynamics, not just a static description.

Vermeij also notes an important difference between the organization of immune systems and organisms and the so-called self-organized systems such as social networks and food webs (Vermeij 2004). Self-organized networks generally lack any deliberate central regulation, although their structure and function are clearly shaped by their environment. Vermeij argues that true intelligence arises in a network only when there is constant feedback between the bottom-up "learning" of the self-organized network components and the top-down regulation and control of a central coordinator. The challenge then for society is to transform security agencies into lightly regulated networks of semiautonomous problem solvers. Notably, Jean-Francois Rischard, a European World Bank vice president, has proposed such a system, based on quickly formed, minimally bureaucratic "global issues networks" for solving international problems from environmental degradation to poverty (Rischard 2002).

By contrast, one of the most high-profile actions of the U.S. government in response to the 9/11 attacks was to create a huge new bureaucracy in the form of a new Department of Homeland Security (DHS). The creation of DHS was touted in part as a response to the perceived lack of coordination and communication among security agencies. Yet a biological viewpoint shows us that molecules, cells, and complex components of organisms communicate effectively, even with limited central control. While the neuronal networks, ionic pathways, and other biological technology that allow such diffuse coordination evolved over long time periods, our rapidly evolving information technology already allows for clear communication among widely dispersed entities. On the issue of communication and responsiveness, DHS failed its first major test. The U.S. Federal Emergency Management Agency (FEMA), which was subsumed into DHS, was strikingly ineffective in response to the devastation of Hurricane Katrina and continues to provide inadequate services in its wake.

In this volume Elizabeth Prescott asserts that where government bureaucracies fail, markets may take up an organization that looks more like biological systems. Yet deliberate effort is needed to make a market entity operate as efficiently, flexibly, and robustly as possible. This follows from the observation that even in an environment of intense selection, there are organisms can that get by with a less-specialized, more centrally controlled organization. The price these organisms pay is that they do not command as much economic power (e.g., ability to control resources) as the more nimbly structured organism (Vermeij 2004). Indeed, much of the success of Internet giant Google has been attributed to a structure in which managers are continually encouraged to develop independent projects and test new ideas, even though many of these have failed (Lashinsky 2006). Likewise, control systems methods used in complex industries such as oil refining rely on a structure that is hierarchical, but in which multiple specialized processes act relatively autonomously toward creating the whole finished product. Interestingly, fisheries scientist William de la Mare (2005) has proposed applying this theory, which mimics biological organization, to the deliberate management of biological ecosystems by resource agencies.

In rare cases where government security agencies have attempted the biological model of organization, they have been remarkably and sometimes shockingly successful. In 2002 the US Defense Advanced Research Projects Agency (DARPA) announced a Grand Challenge to create robotic vehicles that can independently navigate a complex obstacle course. Rather than hire a single contracting firm with a specific design in mind, DARPA presented an open challenge to a diffuse population of civilian groups.

While the first iteration of the Challenge in 2004 was fraught with failure, groups then learned from one another, and independently modified the wide variety of first-generation designs, selecting out poor performers and replicating successful components. The result was that just a year later several entrants completed the course and all but one traveled farther than all of the vehicles of the previous year. The quick development and low cost of this project relative to other military contracts is testament to the power of organization.

In his book *Blink* (2005), Malcolm Gladwell recounts the story of a military exercise designed to test impressive new central informational processing tool, called Operational Net Assessment, by pitting a mock rogue state against the US military in the Persian Gulf. To combat the massive technological and informational superiority of the military, the rogue state commander gave his operatives few rules and broad leeway to devise and carry out operations against enemy infrastructure. As he said, "the overall guidance and the intent were provided by me and the senior leadership, but the forces in the field wouldn't depend on intricate orders coming from the top" (Gladwell 2005). This created an organizational structure that mirrors that of Google, the DARPA challenge, and the organisms that Vermeij studies. To the surprise and consternation of the military side, the rogue commander brought the US military to its knees almost before the military had been able to execute the first orders of central command. Unfortunately, according to Gladwell, rather than taking this defeat as a lesson, the military commanders re-ran the simulation under a new set of rules that effectively eliminated the semiautonomous structure that the rogue commander had set up.

## Behavior

There are two broad ways in which a focus on animal behavior may improve our security. First, we can observe what animals do, as predator and prey, as a lens with which to examine our security strategies. Second, we can look at human behaviors that threaten our security—either directly as the result of violent action against other humans, or indirectly as a result of ineffective or counterproductive responses to a given threat—and uncover the evolutionary root causes of those behaviors.

Uncertainty, which Vermeij has called the greatest challenge to security, is the driving force behind many animal behaviors. Flocking and herding may decrease uncertainty by spreading individual risk over greater numbers. At the same time predators have a higher certainty of getting a meal by initially targeting a group of animals rather than an individual. Prey use color changes and mimicry to increase uncertainty for a potential predator. Predators may lie in wait to ambush prey, creating uncertainty in the environment. Some prey may then probe their enemy's intentions, as squirrels do with snakes, to reduce that uncertainty. In this way, behaviors may enforce a mutual escalation based on creating and mitigating uncertainty.

The potential for escalation is clear when we switch our thinking back from nature to human society. Dan Blumstein notes in this volume that detection signaling is a low cost way to encourage an enemy to switch to a different target. But for politically minded adversaries, the target may not be as substitutable as say, zebra for springbok. As noted by Keohane and Zeckhauser in an ecological analysis of terrorism (Keohane and Zeckhauser 2003), "averting actions and amelioration affect the threat from terror, but not the underlying capacity." A group determined to kill Israeli civilians, for example, will not switch to Indian citizens because their plans are discovered. In these cases, detection signaling may actually increase vulnerability.

This vulnerability arises in two ways. First, signaling detection of a given threat, which usually manifests as a new defensive structure or a new security procedure, reduces uncertainty for the attacker. For example, if security guards in a government building begin checking the trunks of cars entering the garage for explosives, the potential attacker has one less decision to make about where an effective place to hide an explosive would be. Second, signal detection through defensive action may lead to escalation of deadlier armaments as it does in biological organisms. In nature, escalation can occur between competing individuals (such as male fiddler crabs who grow a single large claw that is displayed aggressively to other males during mating rituals) or between predator and prey (such as when snails develop more armored shells as their crab predators develop ever stronger crushers). In the Cold War, antinuclear defenses—from hardened silos to missile defense systems—were curtailed in treaties because the logical consequence of defense mechanisms would be to encourage an attacker to make more potent weapons. In post–Cold War geopolitics, where so-called rogue nations currently lack the resources to build huge nuclear arsenals or large numbers of long-range missiles, the evolutionary logic of the Cold War is turned sideways. Rather than simply build more missiles or high-yield weapons, the escalation is toward new and potentially deadlier ways to deliver those weapons (such as dispersive warheads on a single missile), lower cost (dirty bombs), and stealthier substitutes (bombs in shipping containers).

It has been argued that signaling in these cases makes the defensive population feel safer by giving it a sense that at least something is being done. Here, Blumstein's work on marmots is instructive. He finds that within a given group of marmots, some, which he calls "Nervous Nellies," signal often and often make false alarms, while others ("Cool Hand Lukes") signal only when there is a clear and present danger. Unlike the parable of the boy who cried wolf, however, other marmots do not adapt to ignore the calls of Nervous Nellies but rather spend extra time and energy (i.e., less time foraging) trying to figure out if the call is meaningful. Thus, signaling to a focused imminent threat may provide some benefit, whereas signaling to a diffuse threat with the potential to attack in many different manners has the multiple detriments of encouraging escalation, reducing uncertainty for the attacker, and confusing the defensive population (Blumstein 2004).

The other aspect of the behavior concept in natural security is to uncover the evolutionary roots of our behaviors. Often, we refer to adversaries as "crazy," "madmen," or "irrational," and indeed, behaviors such as Saddam Hussein's blustering in the face of imminent attack seem irrational. Moreover, behaviors such as suicide bombing even seem to defy the evolutionary tenet of self-preservation. On the other side, analysts have begun to question the rationality of devoting such large resources to the relatively low-risk problem of terrorism (Mueller 2005). Yet in considering these behaviors, and especially in thinking about how to beneficially modify them, we need to realize both that they are deeply rooted and that they developed in environments radically different than our own.

Luis Villarreal in this volume speaks to the deep roots of our behaviors, especially those that confer strong group identity and strong beliefs in even the most nonsensical concepts espoused by the group. His life history of these behaviors predates humans by hundreds of millions of years, tracing back to the needs of early prokaryotic organisms for self-identification and prevention of invasion by conflicting genetic messages. This deep history of group behavior, codified through successive major evolutionary advancements (such as complex sensory organs in vertebrates, color vision, and written language), provides a sense of just how engrained associated belief systems can be. Richard Sosis and Candace Alcorta in this volume expound on the global reach and consequences of belief systems, showing how societal norms and characteristics common across religions are of a type that can facilitate and reinforce the operations of terrorist organizations.

Even given the constraints of group identity and belief, our ability to analyze our security situation on an individual basis is compromised by our evolutionary past. As working group member John Tooby and his partner and colleague Leda Cosmides have said, "Our modern skulls house a stone age mind" (1997), meaning that over 99% of our species' history was spent in small, nomadic hunter-gatherer societies and that the few generations we have spent in densely populated industrial society have not been enough for our minds to adapt genetically to the rapidly changed environment. So when we ask questions such as, "Why do we overprioritize the risk of terrorism relative to much more lethal, yet lower-profile threats such as accidental hospital infections and lung disease?" we have to consider that our ability to quantitatively assess risks, which developed almost wholly in the ancestral environment, is totally unprepared to deal with information that can be beamed to us instantly on the whole range of worldly threats to an enormous and dense human population. With this stone age mind, we overestimate the risk of terrorism to ourselves because we have seen the images of people dying due to terrorism. In an ancestral environment, to see someone dying meant that the threat was decidedly close at hand.

Relying only on historical or sociological analysis of modern humans, which covers only 1% or so of the biological history of human life, to understand the actions of terrorists and our own responses is like trying to understand biology by looking at a few dozen domesticated species. Not only is a view focused on industrial age humans temporally biased, it also considers solely what is still a highly anomalous period in our development. It is no wonder, then, that behaviors viewed without evolutionary context appear irrational. But calling something irrational only identifies the problem, it does not get us any closer to solving it.

## Environmental Awareness

Natural organisms live in a risk-filled and uncertain environment. In theory, organisms have three broad options to deal with this risk. They can either avoid it altogether, live with it, or eliminate it. As Vermeij notes, some organisms avoid their risk to an extreme extent (for example, by adapting to a sedentary, low-metabolism life in the deep sea). Yet, as with organisms that do not develop specialized organization, these organisms pay a price by failing to command significant economic power, which further limits their ability to adapt and to be resilient to change (Vermeij 2004).

Organisms live with and minimize risk in several ways. First, an organism needs to perceive and characterize the risks in its environment. Second, the organism must develop technologies or techniques to avoid or mitigate the risk, and this typically occurs through past experience with the risk. Finally, the organism must have a store of resources available for risk management. The millions of life forms we know of have resulted from adaptations that in large part were developed in response to the organisms' needs to live and reproduce in a risk-filled environment.

In some cases, through energy-intensive efforts, organisms are able to eliminate risks in their environment. Ants and primates, for example, engage in occasional acts of "war" on neighboring groups or competitive species, and these may eliminate (albeit for a limited period of time) perceived risks. Crows are known to mob and chase away owls they find nesting near their home range in order to eliminate this predatory risk to the crows' young. Note that the threats targeted by these actions are generally local and well defined. In that sense, even if these are the principle present threats, they are really just components of a larger risk pool in the environment. Collectively, risk has deep evolutionary roots, is globally distributed, and is highly uncertain temporally. Some examples make this clear. A tuna will always be a predator, freezes and less common weather events such as Hurricanes and tsunamis will happen, and earthquakes will occur (my geologist friend's tongue-in-cheek "Stop Plate Tectonics" T-shirt notwithstanding). Thus, while organisms may use considerable resources to mitigate and avoid risks, they do not generally attempt to eliminate *all* risk in the environment. The reason for this is simple, yet paramount: it takes far more energy to try to eliminate risk than to learn to live with it.

Society, it turns out, runs into many of the same constraints as organisms when calculating how to deal with risk. For an entity, such as a nation, that wants to simultaneously be secure and economically prosperous, avoiding all risk as does a deep sea brachiopod is not an option. Indeed, in society it is politically popular to declare all-out wars on the risks that we face. In some cases, there are clear and specific threats that can indeed be eliminated. The elimination of smallpox is an example in which the threat was well characterized, the technology and techniques to eliminate it well understood, and the resources to apply them available. In World War II, although the allies were less certain that their technologies and techniques would be successful, and the relative resource cost was far greater, the threat was still well characterized and specific.

More recently, however, we have declared wars on a host of diffuse threats with poorly understood root causes and few effective countermeasures, such as the wars on drugs, on terrorism and, in ecology, on invasive species. Yet, as with smallpox and Hitler, these threats are referred to as "evils" that must be eradicated at all costs. Chester Crocker, a former U.S. Assistant Secretary of State, has argued that the declaration of a global war on terror exaggerates threats relative to other problems, allows distinct types of threats to be conflated with one another, and diminishes the sense that we are tied in a complex and global way to the sources of threat (and thus require cooperative solutions at a global scale) (Crocker 2005). Notably, ecologist Brendon Larson has raised similar criticisms against the use of war analogies when dealing with invasive species (Larson 2005).

In essence, the attempt to eliminate terrorism, or any globally distributed and deeply rooted risk, is akin to trying to taking the advice on my friend's "Stop Plate Tectonics" T-shirt seriously. It is bound to waste resources because it fails to prioritize specific key risks, fails to understand the strength of underlying causes, and allows almost anything to be called a partial success (and thus justify continued action), even as there is no reasonable strategy for total victory. Natural selection ensures that biological organisms do not waste resources engaged in risk-elimination strategies, rewarding instead those individuals that commit their limited energy to minimizing risk just enough to survive and reproduce. For society, without the benefit of a selective safety valve, it is critical to deliberately establish a balance in resource allocation between completely eliminating and completely avoiding risk.

Ideally, at this balance point a society, or an organism, can respond effectively to immediate dangers while maintaining ability to handle uncertain future threats. Given the difficulty of finding that balance in a highly uncertain environment, it is telling that generalized defenses—which potentially provide protection even from unknown threats—are the norm in evolutionary history (Vermeij 1987). For example, the evolution of thick armor in invertebrates is a general response against a wide variety of predators that alternatively attempt to break or drill holes to kill their hard-shelled prey. A general defense approach is expressed in a sophisticated way in the adaptive immune system, which can quickly recognize, characterize, and appropriately upregulate responses to specific pathogens that could arise from an extremely diverse potential pool. From an epidemiological viewpoint, organisms can respond to a pathogenic threat using a mixed strategy of resistance (e.g., lowering infection rates) and tolerance (reducing the effects of the pathogen on the host), with the balance between the two determined in response to environmental conditions (Restif and Koella 2004).

Unlike a "war" strategy, which implies an optimal solution (the total elimination of the threat), a generalized strategy is likely to never be optimal ("Jack of all trades, master of none"). Yet nature does not especially value optimal solutions, but rather solutions that are just good enough to ensure survival and reproduction (see Vermeij 2004 for good discussions on this). Popper and colleagues from the RAND Corporation in Santa Monica, California, have also converged on this realization in their emerging futures modeling approaches. As they claim, "Our approach is to look not for optimal strategies but for robust ones. A robust strategy performs well when compared with the alternatives across a wide range of plausible futures. It need not be the optimal strategy in any future; it will, however, yield satisfactory outcomes in both easy-to-envision futures and hard-to-anticipate contingencies" (Popper et al. 2005, 69).

## Timing

In the largest sense, time is a key resource for the proliferation of species and diversity of life on Earth today. Despite several mass extinction episodes, there has been an overall increase in the diversity of groups of species and the types of ecological niches they occupy over the course of Earth's history. Environmental change has continually accompanied this march of time, and the rate of change has intensified as greater numbers of species (including humans), inhabiting greater numbers of spaces and habitat types, interact among themselves, each other, and their environment.

Paleontologists, whose job is to study long periods of time, are deeply aware of the power of change. In this volume, for instance, paleobiologist Gregory Dietl focused his criticism of international relations theory on its failure to address the processes of change in the geopolitical arena. Fields of natural history, including ecology, that are more concerned with the here and now, however, have often presented static descriptions of small slices of natural systems, as if change was a fluctuation within a zero-sum field of background noise. An awareness of the effects of global climate change has jarred ecologists into putting our observations and experiments into the context of a continually changing environment. My work has shown that even highly diverse marine communities may be radically different from the community of just a few decades ago due to the effects of climate warming (Sagarin et al. 1999). My colleague Eric Sanford has shown that the power of top "keystone" predators, once thought to be the dominant force in shaping some communities, may be drastically reduced under expected climatic change regimes (Sanford 1999).

Security policy analysts should have heard a similar wake up call, one dictating that our security environment is continually in flux, following the attacks of 9/11 and the destruction of Hurricane Katrina. Terrorism and weather, like living elements of a natural system, evolve with and drive changes in the landscape on which they are adapting. In the immunological world, for example, pathogens may become more virulent as networks of transmission become increasingly interconnected (Boots et al. 2004). This dynamic clearly applies to terrorism, as the Internet and global media have helped spread terrorist messages and have provided new grounds for recruitment. Yet our security systems continue to presuppose a static security environment, or (only slightly better) an environment that has changed from one static state (e.g., "the Cold War") to another ("the post-9/11 world"). The security systems in Washington, D.C., post 9/11 that I discuss in the introduction to this volume were neither radically different from pre-9/11 security, nor did they change through time. Reports by colleagues of mine demonstrating the importance of dynamic coastal wetland systems, as opposed to static levees, to protect coastal cities such as New Orleans have been buried or ignored by FEMA. The ban on most liquids and gels instituted for air travelers following the discovery of a plot to blow up an airliner with a liquid explosive is a particularly egregious example of a static strategy. While costing untold millions of dollars and hours of delays, the ban does nothing to anticipate that a terrorist organization will adapt or escalate its attack plan, nor does it contain within it an ability to counter such adaptation.

Beyond a basic understanding that environments change, we must also understand how they change. In this regard, it is remarkable that phases of change in nature are parsed similarly at nearly all levels of biological complexity. We observe stages of origins, growth, and decline (though the terminology for the stages varies somewhat) in cell lines within the body, in individual organisms, in lineages of species, in populations of organisms, in multispecies communities, and in whole ecosystems. Importantly, dynamics such as the rate of change, the strength of competition, and the level of vulnerability of key components vary somewhat characteristically between these developmental stages. Thus, threats should not only be viewed as continually changing with time, but also as acting in accordance with the stage of their development.

The origins stage, for example, is where elimination of a threat can be conducted most effectively with the least cumulative damage. Ecologists and epidemiologists now exercise extreme vigilance for the first signs of destructive invasive species (e.g., the marine alga *Caulerpa taxifolia* [Boudouresque et al. 1995]), or emerging infectious disease (e.g., H5N1 avian influenza), which can be warded off only in the early stages. Indeed, recent epidemiological models demonstrate that even in a world with a rapid global transportation network, spread of pathogens can be prevented through vigilance and understanding the dynamics of the global network (Hufnagel et al. 2004). Yet the origins stage is also where the threat is least detectable, making it particularly challenging. Because terror groups tend to form through tightly bound friendship and kin relationships (Sageman 2004), the origin of a new terrorist threat may become known only by its first act of terror.

By contrast, in the growth phase, threats may already be too well established to be dealt with effectively. Some of the most destructive invasive species in ecological systems, such as zebra mussels in the U.S. Great Lakes, were not seen as problematic until their populations had exploded to such numbers that eradication was nearly impossible. Likewise, in the United States, public concern about terrorism did not reach a threshold until the 9/11 attacks, despite earlier attacks on U.S. airliners, the World Trade Center, U.S. embassies, and the *U.S.S. Cole* that were indicative of proliferate growth in the capabilities of al Qaeda. By this stage, the resource cost for control may be prohibitive.

Even in the decline stage there is need for vigilance. Life persists in the wake of mass extinctions and the gravest of epidemics (e.g., 1918 influenza pandemic). Given suitable conditions, the period following a collapse may be one of great innovation and exploitation of resources (Gunderson and Holling 2002), where new terrorist organizations and new infectious diseases may arise. Recent calls to focus intensely on protecting human rights and infrastructure development and stabilizing leadership in failed and failing states (Weinstein et al. 2004; Crocker 2005) are supported by an ecological view of security because they implicitly consider bottom-up effects (the emerging resource base for an organism or organization) and the ecological conditions conducive to species origination and expansion.

## Integrating the Four Concepts

It should be clear by now that the strategies that emerge from adherence to these concepts cannot be applied in isolation, but rather they must be viewed as parts of a whole. Natural selection ensures that organisms in nature innately take this holistic view. In society, where we seek and reward those who promise one optimal solution, this is a hard-learned lesson. As Johnson and E. Madin argue in this volume, our failure to act until after disaster strikes is a function of *both* deep-rooted behavioral biases *and* organizational structures that resist change. Likewise, generalized defenses, and the ability to switch from tolerance to resistance strategies, requires an organizational structure that allows rapid response and reallocation of resources. Commanders in Iraq understood fairly quickly how rapidly the security environment was changing due to proliferation and diversification of improvised explosive devices, but they have been hampered in their countermeasure response by the plodding, centrally controlled military development and procurement process (National Public Radio 2005). By contrast, Google Senior Vice President for Business Operations Shona Brown and Kathleen Eisenhardt relate a case study of a company during the dot-com economic boom that was organized every bit as nimbly as Google but failed to properly assess the resources in its environment and ultimately crashed (Brown and Eisenhardt 1998).

Natural organisms are continually tested, thus ensuring that resources are not overprioritized toward a strategy that is out of step with the environment. Vermeij considers each organism that has arisen on Earth as a hypothesis about how to survive and reproduce given a set of environmental conditions. Notably, a scientific hypothesis is never accepted, rather, as scientists we may reject it or fail to reject it. Thus, every extant organism on Earth is just one that has not yet been rejected. As it survives further testing through generations and millions of years, it may become a stronger hypothesis, but it always remains vulnerable to rejection (Vermeij 2004). Likewise, our security solutions, which can be considered hypotheses on how to best keep ourselves safe, should never be taken as accepted practices. Unfortunately, as a society we do very little testing of our own hypotheses. In many cases, in fact, we treat our hypotheses as sacred cows (the U.S. farm subsidies and a military budget that is largely off-limits to congressional discretion are examples). As such our policy making is less a scientific endeavor than it is a matter of faith.

The result is that we are forced to test our hypotheses during periods of conflict. Wars impose their own natural selection, as well as attendant escalation of strategies, as demonstrated by the rapid evolution of axis and allies aircraft in World War II. While there are many positive outcomes of this violent testing (many civilian technologies emerged out of rapid development in wartime) there are many obvious negative externalities (the proliferation of automatic weapons is a particularly deadly example). Ivan Arreguin-Toft additionally argues that many of today's guerrillas have been "socialized" to the effectiveness of guerrilla warfare through exposure to the many insurgencies world wide following World War II (Arreguin-Toft 2001).

Testing through direct conflict (and its attendant externalities) need not be the only pathway to improvement if we both accept our security solutions as testable hypotheses and, critically, provide the ability to modify those hypotheses. High-tech companies routinely employ computer hackers to test their systems. By contrast, the best tests of potential vulnerabilities in our airport security systems have been run by underemployed students who have taken weapons and mock contraband through security screening and even demonstrated that they can board plans with no positive identification check (see, e.g., Soghoian 2006).

If we are to take on the role of developing new security hypotheses and then testing and modifying them, the first step (as it is in any scientific study) is to consider the range of alternatives that are possible. This volume represents several early steps in that process, and it necessarily covers a limited area in the landscape of natural security. Each topic in this book can be a launching point for further discussion, theory, and empirical testing. Yet there are also many relevant topics, such as the role of symbiosis, the primacy of induced versus constitutive defenses in plants, and the maintenance of dominance hierarchies in animals, among others, that do not get proper treatment here. Accordingly, the members of the working group behind this volume continue to meet in person and virtually, and our discussion continues to grow through novel contributions from practitioners in a wide range of fields.

## Life Lessons

Perhaps the most quoted line from the United States' 9/11 Commission Report (National Commission on Terrorist Acts upon the United States 2004) was that 9/11 represented a "failure of imagination." This does seem a particularly pithy description of intelligence failures leading up to that day. Today, by contrast, failure of imagination is by no means our most worrisome problem. Since 9/11, we imagine all sorts of potential catastrophes. From the types of security systems currently in place it would appear, for example, that we worry intensely about what a terrorist might do to a plane with 3.5 ounces of *Aqua Fresh*, that model rocket engines might be stockpiled and turned into an explosive, or that someone might turn the Botanical Gardens in Washington, D.C., into a fertilizer bomb and send a wall of molten glass up Capitol Hill.

Yet as our imagination runs wild, we are still short on effective solutions. And as we conjure the most implausible scenarios, some of the biggest threats to our lives are glaringly ignored. For years it was known that a hurricane of a certain magnitude landing on New Orleans (left vulnerable by its geographic location, misguided engineering projects, and a historic disregard of the protective services that a natural floodplain can provide) would sink the city, yet no effective preventative or response measures were put into place. If we do so poorly with predictable threats, how can we handle the multitude of threats in our environment that cannot be predicted with certainty?

In fact, the situation in New Orleans provides both the question and the answer. There, a natural, resilient system of security was in place due to coastal marshes and the accretion of materials brought from the Mississippi, but it was wrenched away through inflexible engineering and willful ignorance of natural processes. Likewise, our institutions, our relationships, and our behaviors have been disconnected from their natural origins. It can easily be argued that we have reduced suffering and benefited greatly by moving away from this ancestral state, but that does not mean we should forget its imprint on us, nor its lessons. Paradoxically, our current state of insecurity and uncertainty is both a function of evolution's continued presence in our lives and our freedom from evolutionary constraint.

Even as the security landscape is continually changing, our security systems and strategies often regress to approaches from an era long past. As Popper and colleagues note in regard to futures planning (2005), "in the presence of such deep uncertainty, the machinery of prediction and decision making seizes up. Traditional analytical approaches gravitate to the well-understood parts of the challenge and shy away from the rest" (p. 68). On the other hand, the history of life provides lessons and models for dealing with tenacious enemies, probabilistic solutions where there is no absolute winner, and globally ubiquitous threats, all areas where traditional approaches are often inadequate. Yet nature's secrets are not classified, making them freely available to those who want to adopt them for better or for worse. Thus, it is both sobering and compelling to recall that among the millions of life forms that have inhabited the Earth and engaged in potentially lethal conflicts, one simple rule applies: those who have mastered the evolutionary game survive and prosper—those who have not become extinct.

REFERENCES

Arreguin-Toft, I. 2001. How the weak win wars: A theory of asymmetric conflict. *International Security* 26 (1): 93–128.

Barabasi, A.-L. 2003. *Linked.* London: Penguin Books.

Blumstein, D. T., L. Verneyre, and J. C. Daniel 2004. Reliability and the adaptive utility of discrimination among alarm callers. *Proceedings of the Royal Society of London B.* 271: 1851–1857.

Boots, M., P. J. Hudson, and A. Sasaki. 2004. Large shifts in pathogen virulence relate to host population structure. *Science* 303: 842–844.

Boudouresque, C. F., A. Meinesz, M. A. Ribera, and E. Ballesteros. 1995. Spread of the green alga *Caulerpa taxifolia* (Caulerpales, Chlorophyta) in the Mediterranean: Possible consequences of a major ecological event. *Scientia Marina* 59: 21–29.

Brown, S. L., and K. M. Eisenhardt. 1998. *Competing on the edge: Strategy as structured chaos.* Cambridge, MA: Harvard Business School Press.

Crocker, C. A. 2005. A dubious template for U.S. foreign policy. *Survival* 47 (1): 51–70.

de la Mare, W. K. 2005. Marine ecosystem-based management as a hierarchical control system. *Marine Policy* 29: 57–68.

Diamond, J. 1999. *Guns, germs and steel: The fates of human societies.* New York: W. W. Norton and Company.

Dobzhansky, T. 1973. Nothing in biology makes sense except in light of evolution. *American Biology Teacher* 35:125–129.

Ehrlich, P. R., and S. A. Levin 2005. The evolution of norms. *PLoS Biology* 3 (6): e194.

Gladwell, M. 2005. *Blink. The power of thinking without thinking.* New York: Little Brown and Company.

Gunderson, L. H., and C. S. Holling, eds. 2002. *Panarchy: Understanding transformations in human and natural systems.* Washington, DC: Island Press.

Hufnagel, L., D. Brockmann, and T. Geisel. 2004. Forecast and control of epidemics in a globalized world. *Proceedings of the National Academy of Sciences USA* 101: 15124–15129.

Keohane, N. O., and R. J. Zeckhauser. 2003. The ecology of terror defense. *Journal of Risk and Uncertainty* 26 (2/3): 201–229.

Larson, B. M. 2005. The war of the roses: Demilitarizing invasion biology. *Frontiers in Ecology and the Environment* 3: 495–500.

Lashinsky, A. 2006. Chaos by design. The inside story of disorder, disarray, and uncertainty at Google. And why it's all part of the plan. (They hope.) *Fortune.* Oct. 2, 2006 Online Edition.

Levin, S. A. 1999. *Fragile dominion.* Cambridge, MA: Perseus Publishing.

Mueller, J. 2005. Simplicity and spook: Terrorism and the dynamics of threat exaggeration. *International Studies Perspectives* 6: 208–234.

National Commission on Terrorist Acts Upon the United States 2004. *The 9/11 Commission Report.* Washington, DC: US Government Printing Office.

National Public Radio. 2005. Interview: U.S. Army Brigadier General Joseph Votel and U.S. soldiers describe looking for, finding and destroying IEDs in Iraq, ed. R. Montagne and S. Inskeep, May 20, 2005.

Popper, S. W., R. J. Lempert, and S. C. Bankes. 2005. Shaping the future. *Scientific American* April, 66–71.

Restif, O., and J. C. Koella. 2004. Concurrent evolution of resistance and tolerance to pathogens. *American Naturalist* 164 (4): E90–E102.

Rischard, J. F. 2002. Global issues networks: Desperate times deserve innovative measures. *Washington Quarterly* 26 (1): 17–33.

Sagarin, R. D., J. P. Barry, S. E. Gilman, and C. H. Baxter. 1999. Climate-related changes in an intertidal community over short and long time scales. *Ecological Monographs* 69 (4): 465–490.

Sageman, M. 2004. *Understanding terror networks.* Philadelphia: University of Pennsylvania Press.

Sanford, E. 1999. Regulation of keystone predation by small changes in ocean temperature. *Science* 283: 2095–2097.

Sapolsky, R. M. 2006. A natural history of peace. *Foreign Affairs* 85 (1) Online Edition.

Soghoian, C. 2006. Slight paranoia [blog]. Available at http://slightparanoia. blogspot.com/2006_09_01_slightparanoia_archive.html.

Tooby, J., and L. Cosmides. 1997. Evolutionary psychology: A primer. Center for Evolutionary Psychology, University of California, Santa Barbara. Available at www. psych.ucsb.edu/research/cep/primer.html.

Vermeij, G. J. 1987. *Evolution and escalation: An ecological history of life.*, Princeton, NJ: Princeton University Press.

Vermeij, G. 2004. *Nature: An economic history.* Princeton: Princeton University Press.

Weinstein, J. M., J. E. Porter, and S. E. Eizenstat 2004. *On the brink: Weak states and U.S. national security.* Washington, DC: Center for Global Development.

# INDEX

Abdo, Hussam, 131
Abdullah, King of Jordan, 62
active defense/engagement, 30, 31–32, 38, 39, 229
adaptation, 25–30, 34, 35, 225–26; and attention, 225; characteristics of states promoting, 225; in ecosystems, 249; and extinctions, 30, 38, 181; and government, 16–17; and human nature, 26–27; of immune system, 229–30; and innovation, 225, 228, 231; to risk (*See* risk, adaptation to); security and unpredictability in, 25–26, 28–29, 31, 36, 37, 39, 228–29. *See also* disaster and change; evolution
adaptive immunity, 49–52; defined, 63. *See also* group identity and immunity; immune systems
addiction modules, 42, 43, 45–48, 51, 56, 61; and pair-bonding, 59–60
adolescence, 61, 113; and terrorism, 115–16, 118, 120, 131, 144
Afghanistan, 134, 164, 228; Taliban in, 127
afterlife rewards, for suicide terrorists, 108, 112–13, 114, 129, 131–32, 197
alarm signals, 149–50, 152–53. *See also* signaling theory
Alcorta, Candace S., 9, 105, 109, 113, 268
altruism, 46, 48
anarchy, in international politics, 127
anthrax attack, 17, 82
antipredator behavior, 147–57; and adaptation, 31, 32, 34, 36, 37–38; and alarm signals, 149–50, 152–53; avoiding costly behaviors, 150–51, 154; defense strategy, 149–50, 151, 154; and DHS,

152–53, 155–57; direct *vs.* indirect risk cues, 148; flexible responses, 152, 153; future research needs, 154–55; and habituation, 152, 153, 155, 157; and multipredator hypothesis, 151–52; overestimating risk in, 148–49, 153; policy implications, 152–54. *See also* predator-prey model
apoptosis (programmed cell death), 44, 46, 48, 49, 50, 60–61; defined, 63
Arabian Peninsula, 129. *See also* Saudi Arabia
Argo, N., 117
arms race, 160; in nature, 9, 36, 37–38
Arreguin-Toft, Ivan, 275
al-Askari mosque bombing (2006), 166, 169
Atran, Scott, 9, 109, 117, 119, 120, 141–44
audio learning, 56, 57–58
authority, 35
avian influenza ($H_5N_1$), 63, 73–74, 79, 195–96, 201, 273
Azam, J.-P., 112
Aziz Rantisi, Abdul, 107

Bambach, R. K., 38
Barabasi, A.-L., 264
behavior, and security, 266–69. *See also* predator-prey model
behavioral ecology, 8*t*, 12, 147, 148, 153
belief, 11, 60–61, 110, 268; and addiction modules, 61; and cognitive immunity, 58, 62; and indoctrination, 132–33; moral belief, 9, 107–8, 113, 141–44. *See also* religion and terrorism
Benjamin, D., 105
Bering, J., 114